FIGURE OF FRONT COVER

Diagrammatic representation of velocity profiles while a drilling fluid is circulating through an annulus. The profiles depend on annulus geometry, flow rate and rheological characteristics (determined on field with a Fann viscometer).

AMONG OUR BOOKS

**PUBLICATIONS DE LA CHAMBRE SYNDICALE
DE LA RECHERCHE ET DE LA PRODUCTION DU PÉTROLE ET DU GAZ NATUREL**

Publications in English

- Evaporite Deposits. Illustration and Interpretation
  of some Environmental Sequences.

- Blowout Prevention and Well Control.

chambre syndicale de la
recherche et de la production
du pétrole et du gaz naturel

comité des techniciens

commission exploitation
sous-commission laboratoires d'exploitation
groupe fluides de forage et ciments

DRILLING MUD AND CEMENT SLURRY RHEOLOGY MANUAL

1982

ÉDITIONS TECHNIP 27 RUE GINOUX 75737 PARIS CEDEX 15 techni**p**

ISBN 978-94-010-9248-7   ISBN 978-94-010-9246-3 (eBook)
DOI 10.1007/978-94-010-9246-3

Preface

The importance of rheological properties — i.e., of flow properties — of drilling fluids and cement slurries is now considered as self-evident.

During the drilling, the control of the "viscosity", or the rheological characteristics, of the mud by simple and rapid methods has always been an indispensable guide for the technicians on the worksite, and has remained so to this day. In fact, flow properties directly affect the progress of the operations. Their proper adjustment ensures that the hole will be to gauge and that it will be drilled in the shortest possible time and under the best possible safety conditions.

In a similar manner, the efficiency with which the mud is displaced by cement slurry during cementing depends on the flow properties of the fluids involved.

Attempts at adjustment and optimization of rheological characteristics within the framework of other constraints may obviously involve laborious calculations. However, in recent years the treatment of the data has become much easier due to the use of modern computing techniques, in particular with programmable mini-computers.

The group "Drilling Fluids and Cements" of the French Technical Committee of the *Chambre Syndicale de la Recherche et de la Production du Pétrole et du Gaz Naturel* wanted to compile data on rheology of fluids in a single manual, to be used by members of the profession.

This book is not primarily concerned with theory, regardless of the numerous equations it contains at the risk of discouraging the reader; its basic approach is practical. It has attempted to present a logical treatment which will be easy to apply in practice. As a result, certain computing methods were omitted, and precision sometimes had to be sacrificed to simplicity. However, no apology is made for the use of such approximations; in fact, any attempt at rigor would be doomed to failure, in view of the many inherent factors which do not lend themselves to quantitative treatment.

Chapter 1 deals with fundamental concepts.

Chapter 2 refers to the general principles involved in determining rheological parameters of drilling fluids and cement slurries.

Chapter 3 relates to practical methods for using the results obtained in the first two Chapters, in units employed on the worksite. It is primarily intended for technicians called upon to make "hydraulic" computations during drilling.

Chapter 4 contains several examples.

It is both my duty and my pleasure to make the following acknowledgements.

This book has been edited by the Group "Drilling Fluids and Cements", of the French *Chambre Syndicale de la Recherche et de la Production du Pétrole et du Gaz Naturel*, Exploitation Commission, Exploitation Laboratories Sub-Commission, headed by Raymond Broc (*Compagnie Française des Pétroles*).

The original idea of writing this manual was that of Jacques de Lautrec (*Société Nationale Elf Aquitaine-Production*). Its realization would have been impossible without the assistance, collaboration, and tireless efforts of members and highly qualified technicians of different organizations, of whom special mention should be made:

CECA (SA)

Compagnie Française des Pétroles (CFP)

IMCO Services

Institut Français du Pétrole (IFP)

Société Nationale Elf Aquitaine-Production (SNEA-P)

Madeleine Martin (*IFP*) coordinated the individual activities.

I wish to thank them all for their valuable contributions.

R. Monicard
President, Laboratory Sub-Commission
of the Exploitation Commission

Contents

CONTENTS

2

APPLICATION TO DRILLING FLUIDS AND CEMENT SLURRIES

3

PRINCIPAL METHODS OF EVALUATION

4

PRACTICAL EXAMPLES

Symbols and Units

In order to facilitate the understanding of Chapters 1 and 2, all equations given in these Chapters are in SI system of units. A supplementary list of symbols and units used in Chapters 3 and 4, and a conversion Table, will be found on p. 72 ff.

Symbols	Meaning	Dimensions	Units SI
A	Total nozzle surface area	L^2	m^2
Av	Drilling rate	LT^{-1}	m/s
C_r	Drag coefficient (settling)	none	
D	Inner diameter of drill string	L	m
D_i	Inner diameter of annulus (outer diameter of drill string)	L	m
D_o	Outer diameter of annulus	L	m
f	Friction (flow) factor	none	
g	Acceleration due to gravity	LT^{-2}	m/s^2 (9.81)
g_t	Gel strength at time t	$ML^{-1}T^{-2}$	Pa
I_h	Hydraulic impact	MLT^{-2}	N (Newton)
K	Consistency index	$ML^{-1}T^{n-2}$	$Pa \cdot s^n$
L	Length	L	m
n	Power-law index of flow behavior	none	
P	Pressure	$ML^{-1}T^{-2}$	Pa
P_h	Hydrostatic pressure	$ML^{-1}T^{-2}$	Pa
P_i	Discharge pressure	$ML^{-1}T^{-2}$	Pa
P_s	Service pressure of mud circuit	$ML^{-1}T^{-2}$	Pa
$\mathscr{P}$	Power	ML^2T^{-3}	W (watt)
$\mathscr{P}_h$	Hydraulic power	ML^2T^{-3}	W
$\mathscr{P}_{hc}$	Hydraulic power in circuit outside the bit	ML^2T^{-3}	W
$\mathscr{P}_{he}$	Hydraulic power at the bit	ML^2T^{-3}	W
$\mathscr{P}_{hl}$	Maximum available hydraulic power	ML^2T^{-3}	W

Symbols	Meaning	Dimensions	Units SI
Q	Flow rate	L^3T^{-1}	m³/s
Q_c	Critical flow rate	L^3T^{-1}	m³/s
Q_m	Minimum flow rate ensuring uplift of cuttings in the annulus	L^3T^{-1}	m³/s
Q_s	Maximum flow rate at pressure P_s	L^3T^{-1}	m³/s
Re	Reynolds number	none	
Re_c	Critical Reynolds number	none	
Re_s	Reynolds number for fall of a single particle	none	
v	Volume of one cutting particle	L^3	m³
V	Fluid velocity	LT^{-1}	m/s
V_b	Mud annular velocity	LT^{-1}	m/s
V_c	Critical fluid velocity	LT^{-1}	m/s
V_r	Uplift velocity of cuttings	LT^{-1}	m/s
V_s	Slip velocity of cuttings	LT^{-1}	m/s
$\dot{\gamma}$	Shear rate	T^{-1}	1/s
ΔP	Pressure losses	$ML^{-1}T^{-2}$	Pa
ΔP_a	Pressure losses through the annulus	$ML^{-1}T^{-2}$	Pa
ΔP_c	Pressure losses through the circuit outside the bit nozzles	$ML^{-1}T^{-2}$	Pa
ΔP_e	Pressure losses through the bit	$ML^{-1}T^{-2}$	Pa
ΔP_l	Maximum permissible pressure losses through the circuit	$ML^{-1}T^{-2}$	Pa
μ	Dynamic viscosity	$ML^{-1}T^{-1}$	Pa . s
μ_a	Apparent viscosity	$ML^{-1}T^{-1}$	Pa . s
μ_e	Equivalent viscosity	$ML^{-1}T^{-1}$	Pa . s
μ_p	Plastic viscosity	$ML^{-1}T^{-1}$	Pa . s
ρ	Density	ML^{-3}	kg/m³
ρ_l	Density of fluid	ML^{-3}	kg/m³
ρ_s	Density of solid particle	ML^{-3}	kg/m³
τ	Shear stress	$ML^{-1}T^{-2}$	Pa
τ_0	Yield-point (limiting shear-stress)	$ML^{-1}T^{-2}$	Pa
φ_e	Equivalent diameter of annulus	L	m
φ_h	Hydraulic diameter	L	m
φ_p	Equivalent diameter of the cutting	L	m

1
Fundamental Concepts

1.1. DEFINITION OF RHEOLOGY

Rheology is the science of deformation of materials (if they are solid), or of their flow (if they are liquid) under applied stress.

A force applied to a body produces a deformation in it. In the case of a solid, this deformation will be elastic if the body reverts to its original state as soon as the force is removed (e.g., stretching of a rubber band); it will be plastic if it reverts to its original state only under the action of other forces (e.g., deformation of a ball of putty). In the case of a fluid, applied force induces flow.

1.2. TYPES OF FLOW

There are various types of flow.

1.2.1. Steady flow

The flow is continuous, and may be of the following types:

(a) **Laminar flow,** in which any laminar layer of the fluid is displaced, with respect to other laminar layers, in parallel to the direction of flow, and is moving at its individual speed. As shown by Figure 1, for flow through a cylindrical tube, the flow rate is highest along the axis of the tube. At the tube wall it is zero in the absence of slippage.

(b) **Turbulent flow,** in which small eddies are formed throughout the volume of the fluid (see Fig. 2).

(c) **Plug flow,** in which the fluid moves along the tube as though it were a plug. The flow rate in a given direction is constant, whatever the distance from the axis of the tube may be (see Fig. 3).

Plug flow is displayed by only a few types of fluids. A bentonite suspension, which is a plastic fluid, can display plug flow — unlike water, diesel oil, glycerol solution, etc.

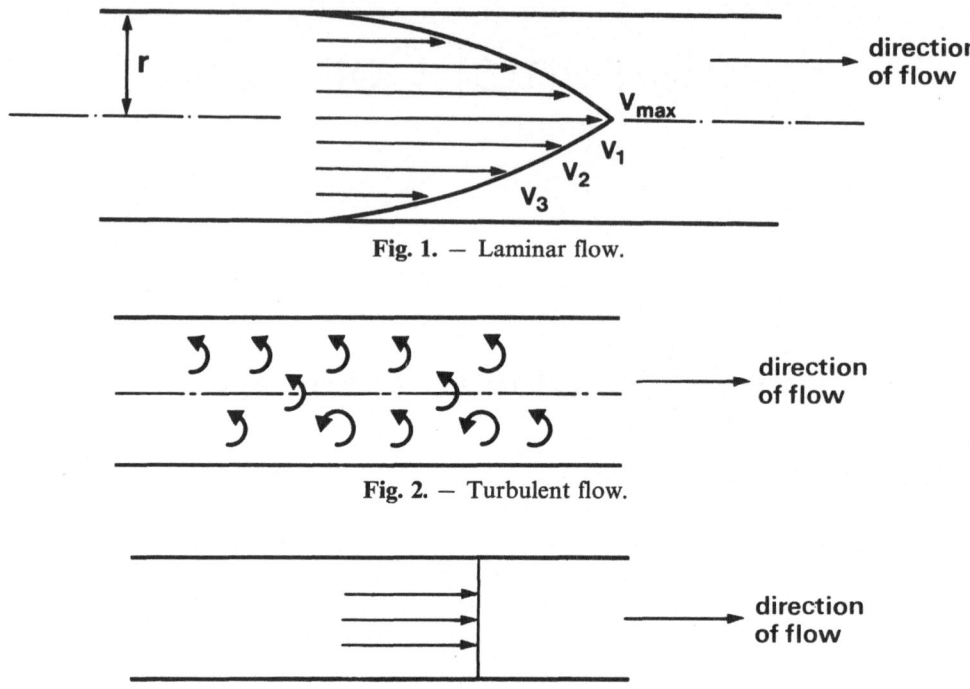

Fig. 1. — Laminar flow.

Fig. 2. — Turbulent flow.

Fig. 3. — Plug flow.

1.2.2. Transient flow

Transient flow occurs just after the conditions of flow have been modified and before the new, permanent conditions of flow have become established.

This condition may occur, for example, at the moment the fluid flow is starting, when the flow rate changes, when the conduit cross-section becomes wider or narrower, etc.

1.2.3   Change of flow type with the average rate of flow

Depending on the average flow velocity V, the following situations may occur for a given type of fluid flowing through a straight cylindrical conduit of given dimensions:

(a) **In the absence of plug flow,** as illustrated in Figure 4.
(b) **In the presence of plug flow,** as illustrated in Figure 5.

Here, V_1 is the critical velocity for termination of laminar flow, and V_2 is the critical velocity for incipient turbulent flow.

NOTE.   The meaning of "transient flow" is broader than that of "laminar-turbulent flow in the transition zone".

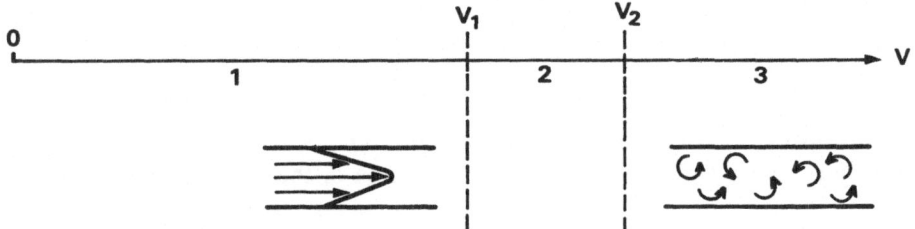

Fig. 4. — Change of flow type.

1. Laminar flow (steady).
2. Laminar-turbulent flow zone (transient).
3. Turbulent flow (steady, on the average).

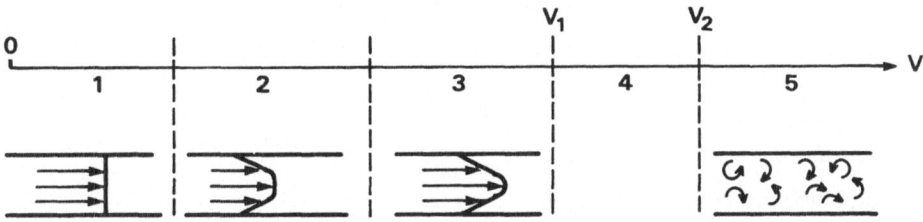

Fig. 5. — Change of flow type.

1. Plug flow (steady).
2. Core thickness decreases with increasing V (transient).
3. Laminar flow (steady).
4. Laminar-turbulent flow (transition zone).
5. Turbulent flow (steady, on the average).

1.3. RHEOLOGICAL CHARACTERISTICS

At a given temperature and pressure, fluids are characterized by:

A. Their behavior under transient conditions, as manifested by their **response time** to changed conditions of flow.

B. Their behavior in laminar flow, characterized by their **experimental flow curve,** or **rheogram.** The constant coefficients of the **equation of flow** represented by this curve are **rheological parameters,** specific to the particular fluid.

If the flow is laminar, the equation of flow relates the shear stress τ with the shear rate $\dot{\gamma}$. For a given fluid, it varies with temperature and pressure.

We have said that in laminar flow the fluid is sheared into laminar layers, parallel to the direction of flow, each layer moving at its specific velocity. We may accordingly define:

(a) **A shear rate** such that

$$\dot{\gamma} = \frac{dV}{dr} = \frac{\text{Velocity difference between two adjacent layers}}{\text{distance between the two layers}} \tag{1}$$

The dimensional equation of $\dot{\gamma}$ is

$$\frac{LT^{-1}}{L} = T^{-1}$$

i.e., the dimension of $\dot{\gamma}$ is an inverse time (s^{-1} or 1/s).

(b) **A shear stress,** which is the force per unit area of the laminar layer inducing the shear.

The dimensional equation of τ is

$$\frac{MLT^{-2}}{L^{-2}} = ML^{-1}T^{-2}$$

i.e., τ has the dimensions of pressure. It is often expressed in lb/100 ft^2 (lb force/100 ft^2) or, in the International System of Units (SI) in pascal (Pa).

For a given shear rate, the **apparent viscosity** μ_a is defined by the equation:

$$\mu_a = \frac{\tau}{\dot{\gamma}} \tag{2}$$

where τ is the shear stress leading to $\dot{\gamma}$.

The dimensional equation of μ_a is

$$\frac{ML^{-1}T^{-2}}{T^{-1}} = ML^{-1}T^{-1}$$

i.e., μ_a has the dimensions of viscosity.

In the SI system, μ_a is expressed in pascal-seconde (Pa . s). The unit which is usually employed is its sub-multiple — the millipascal-seconde (mPa . s). It is equal to the centipoise (cP).

It is often necessary to consider the shear stress, shear rate and apparent viscosity at the wall of the channel where the fluid is moving.

C. Their behavior at rest, as manifested by gel formation after a certain period of time, for **thixotropic fluids.**

A fluid is thixotropic if:

(a) It forms a gel after being shaken and left to stand.
(b) It returns to its original condition after it has been shaken again.

At constant temperature and pressure, thixotropic behavior is **reversible.**

1.4. EXPERIMENTAL DETERMINATION OF RHEOLOGICAL PARAMETERS

The following instruments are used in field practice:

(a) The Marsh funnel viscometer.
(b) The Fann viscometer.

1.4.1. The Marsh funnel viscometer (Fig. 6)

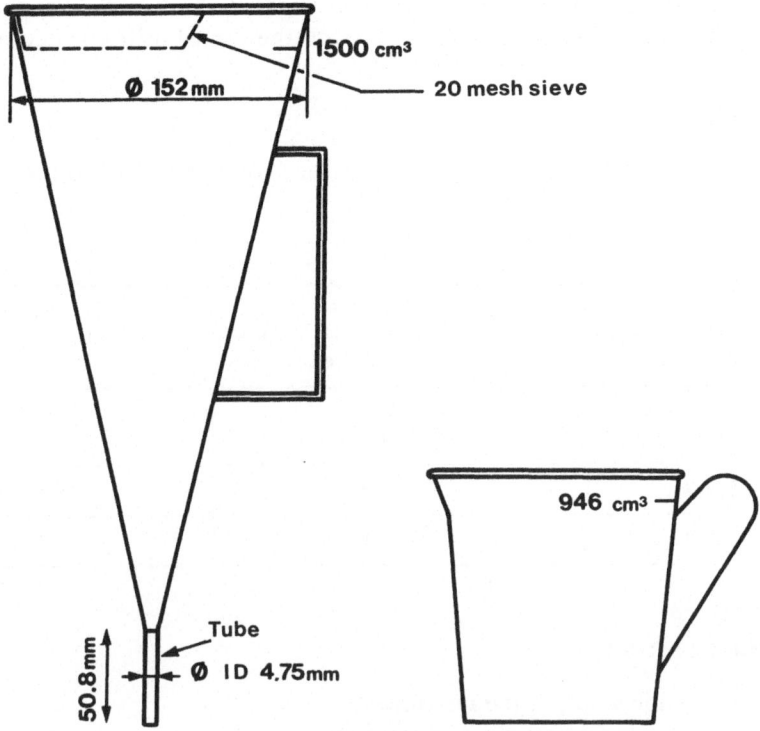

Fig. 6. — Marsh funnel viscometer.

1.4.1.1. Principle

This is a static-type instrument. The variable measured is the time, in seconds, required for a given quantity of the mud to pass through the tube of the instrument, the latter being simply a standardised funnel. The mud is collected in a graduated cup.

1.4.1.2. Procedure

Take the funnel in your hand and stop the orifice tip with your finger.

Pour the mud through the sieve of the funnel, until the mud level is flush with the sieve; this corresponds to a volume of 1 500 cm³.

Holding the funnel by the handle, press a stopwatch and let the mud flow into a graduated cup. Press the stopwatch again when 946 cm³ (1/4 of a gallon) are collected in the cup. The number of seconds on the stopwatch is the Marsh viscosity of the mud.

NOTE.   In other variants of the method, 1 000 cm³ or 500 cm³ of the mud are collected, or else the funnel is filled with 500 cm³ of the mud, and the time required to empty the funnel is determined. However, the **standard** procedure is as given above. If another variant is employed, the volume of the mud poured into the funnel and the volume of the mud collected in the cup must be reported.

Example: "1 500 cm³/500 cm³" if the funnel is filled to the level of the sieve, and the time required for 500 cm³ of the mud to flow through the funnel orifice is determined.

1.4.1.3. Calibration

The instrument is calibrated with pure water. The times thus found are shown in Table 1.

TABLE I

VOLUME OF WATER		TIME OF EFFLUX (s at 20 ± 1° C)
cm³ in funnel	cm³ collected	
1 500	946 (1/4 gallon)	26 ± 1
1 500	1 000	28 ± 1
1 500	500	14

1.4.1.4. Error sources

(a) Erroneous calibration of the instrument.

(b) Orifice tip blocked by caked or imperfectly sieved mud.

(c) Too much time was allowed to pass from the moment the funnel was full and the initial moment of efflux (the mud may have gelled, thus extending the efflux time).

(d) Efflux time inaccurately determined; a stopwatch **must** be used.

1.4.2. The Fann viscometer (Fig. 7)

1.4.2.1. Principle

This instrument is of the rotational coaxial-cylinder type. The common commercial models have:

(a) Either two rotation speeds (600 and 300 rpm). The rotor may be operated manually (*hand-crank* viscometer) or by an electric motor (*motor-driven* viscometer).

(b) Or six rotation speeds (600, 300, 200, 100, 6 and 3 rpm). The rotor is driven by an electric motor.

The dimensions of the rotor sleeve and bob are specified in *API* RP 13B.

The shear stress (scale reading) is determined as a function of the shear rate (from the speed of rotation).

Fig. 7. – Fann viscometer model 35 (Photo *IFP*).

1.4.2.2. Equations to be used with rotational coaxial-cylinder viscometers

We shall introduce the following symbols:

r_i = radius of the inner cylinder,
r_e = radius of the outer cylinder,
h = height of the immersed cylinder,
C = torque acting on the inner cylinder,
Ω = angular velocity of the rotor, defined as

$$\Omega = \frac{2\pi N}{60}$$

where N is the rotation speed in rpm.

The shear stress at the inner cylinder wall is given by

$$\tau_i = \frac{C}{2\pi r_i^2(h + \varepsilon)} \tag{3}$$

where ε is a correction term, having dimension of length, introduced to allow for end-effects.

The shear rate at the inner cylinder wall is

$$\dot{\gamma}_i = \frac{2\Omega}{1 - \left(\dfrac{r_i}{r_e}\right)^2} = \frac{4\pi N}{60\left[1 - \left(\dfrac{r_i}{r_e}\right)^2\right]} \tag{4}$$

if $\dfrac{r_i}{r_e} > 0.9$.

1.4.2.3. Application to Fann viscometer

The instrument has the dimensions:

$$r_e = 18.42 \text{ mm}$$
$$r_i = 17.25 \text{ mm}$$
$$h = 38. \quad \text{mm}$$

The torque, proportional to the scale deflection, is given by

$$C = k\theta$$

where θ is the scale reading and k the spring constant.

For most springs

$$k = 3.87 \ 10^{-5} \text{ N . m/scale unit}$$

We shall now derive the general equations:

Shear stress τ_i
  (a) in SI units (τ_i in pascal):

$$\tau_i = \frac{k\theta}{2\pi r_i^2(h + \varepsilon)} = 0.51\theta \tag{5}$$

  (b) in american units (τ_i in lb force/100 ft^2):

$$\tau_i = \theta \tag{5 bis}$$

for then

$$\frac{k}{2\pi r_i^2(h + \varepsilon)} = 1$$

Shear rate $\dot{\gamma}_i$

$$\dot{\gamma}_i = \frac{4\pi N}{60\left[1 - \left(\dfrac{r_i}{r_e}\right)^2\right]} = 1.7N \tag{6}$$

The Table shows the relation between N and $\dot{\gamma}_i$:

N (rpm)	600	300	200	100	6	3
$\dot{\gamma}_i$ (s^{-1})	1 020	510	340	170	10	5

1.4.2.4. Procedure

The mud is passed through the screen of a Marsh funnel, mechanically stirred for 5 min, and is then poured into the container of the viscometer.

The coaxial cylinders are immersed (the instrument is provided with a slide bar and a threaded clamp) until the reference mark on the rotor is flush with the mud surface.

The rotor is run at 600 rpm (by appropriate setting of the velocity selector and of the switch) and the scale deflection is read.

Without stopping the rotor, the rotation speed is reset to 300 rpm, and the scale deflection is read again.

This routine is repeated for other rotation speeds.

NOTES.

(a) The viscometer requires practically no calibration. If it is nevertheless desired to calibrate the instrument, it should be run with aqueous glycerol solutions of known viscosities, and the flow curve plotted in Cartesian coordinates, with the rotation speeds plotted as abscissa, and scale deflections, as ordinates. The resulting graph is a straight line passing through the origin, as glycerol solutions are Newtonian liquids.

(b) The rotor and the stator are so designed that kinetic energy effects are eliminated, and the density of the sample does not therefore affect the results.

1.4.2.5. Determination of apparent viscosity

The apparent viscosity, μ_a, of drilling fluids and cement slurries is to be determined, in accordance with *API* standard, at a shear rate of 1 020 s^{-1}, which corresponds to a Fann viscometer rotor-speed of 600 rpm. Most often, these conditions are not explicitly stated, but are implied in the term "apparent viscosity".

If μ_a is expressed in pascal-seconde, we obtain, by combining Eqs. 2, 5 and 6:

$$\mu_a = \frac{\tau}{\dot{\gamma}} = \frac{0.51\theta_{600}}{1\,020} \tag{7}$$

If μ_a is expressed in centipoise, we have:

$$\mu_a = 0.5\theta_{600} = \frac{\text{Fann reading at 600 rpm}}{2} \tag{8}$$

1.4.2.6. Determination of thixotropy

Thixotropy can be estimated by observing the changes in strength taking place in a gel as function of time.

According to *API* RP 13 B, two values — the 10 second gel-strength (g_0) and the 10 minute gel-strength (g_{10}) — are determined in a two-speed Fann viscometer. The following procedure is employed. The rotor is run at 600 rpm during 30 s after which the motor is stopped and the small upper knurled knob is set in its intermediate position.

After 10 s have elapsed, the large knurled knob on top of the instrument is slowly rotated counterclockwise by hand, at about 3 rpm.

The maximum scale deflection is noted. This figure is the initial gel-strength in lb/100 ft².

A period of 10 min is allowed to elapse without disturbing the mud, and the operation is repeated. The maximum scale deflection is equal to the 10 minute gel-strength in lb/100 ft².

If the six-speed Fann viscometer is used, the procedure is similar, but the 3 rpm speed is used instead of turning the rotor manually.

1.4.2.7. Comparison between the flow in a rotational coaxial-cylinder viscometer and in a straight cylindrical pipe

A synopsis of the equations for the calculation of τ and $\dot{\gamma}$ for flows taking place in a coaxial-cylinder viscometer and in a cylindrical pipe respectively, will be found in Table II.

TABLE II

ROTATIONAL COAXIAL-CYLINDER VISCOMETER		CYLINDRICAL TUBE	
Inner cylinder radius	$= r_i$	Tube length	$= L$
Outside cylinder radius	$= r_e$	Tube radius	$= r$
Height of cylinder immersion	$= h$		
Torque acting on inner cylinder	$= C$	Pressure loss between two ends of tube	$= \Delta P$
Shear stress at inner cylinder wall	$= \tau_i$	Shear stress at tube wall	$= \tau_r$
$$\tau_i = \frac{C}{2\pi r_i^2 h} \qquad (3)$$		$$\tau_r = \frac{r\Delta P}{2L} \qquad (9)$$	
Angular rotation rate	$= \Omega$	Flow rate Q or average speed:	
		$$V = \frac{Q}{\pi r^2} \qquad (10)$$	
$\downarrow$		$\downarrow$	
Shear rate at wall of inner cylinder $\dot{\gamma}_i$		Shear rate at wall of tube $\dot{\gamma}_r$	
$$\dot{\gamma}_i = \frac{2\Omega}{1 - \left(\dfrac{r_i}{r_e}\right)^2} = \frac{4\pi N'}{1 - \left(\dfrac{r_i}{r_e}\right)^2} \qquad (4)$$		$$\dot{\gamma}_r = \frac{4Q}{\pi r^3}\frac{3n+1}{4n} = \frac{4V}{r}\frac{3n+1}{4n} \qquad (11)$$	
$\left(\text{if } \dfrac{r_i}{r_e} > 0.9\right)$		where	
		$$n = \frac{\text{d}(\log \Delta P)}{\text{d}(\log Q)} \qquad (12)$$	
N' is the number of rounds/seconde			

1.5. RHEOLOGICAL CLASSIFICATION

Table III shows a simplified rheological classification based on the criteria discussed in section 1.3.

TABLE III

A SIMPLIFIED RHEOLOGICAL CLASSIFICATION

FLOW \ FLUID	VISCOUS		VISCOELASTIC
Transient	Response time = 0		Response time ≠ 0
Laminar	Newtonian	Non-Newtonian	Non-Newtonian
Turbulent			Decrease in pressure losses ([1])
Rest	Non thixotropic	Thixotropic or non thixotropic	

([1]) With certain viscoelastic fluids, pressure losses are lower than they would be in water circulating at the same rate in the same pipe (e.g., certain polymer muds).

1.6. RHEOLOGICAL EQUATIONS

1.6.1. Newtonian fluids

The shear stress of Newtonian fluids is directly proportional to the shear rate: if one variable is doubled, the other one is doubled also. The rheological equation is

$$\tau = \mu\dot{\gamma} \tag{13}$$

The following plot is obtained in Cartesian coordinates (Fig. 8).

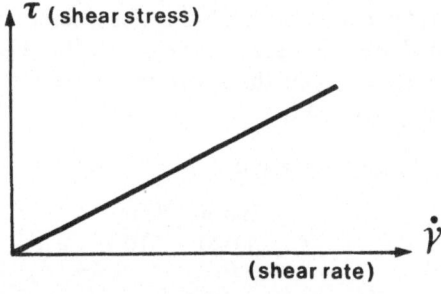

Fig. 8

The graph is a straight line passing through the origin; the fluid begins to move as soon as a nonzero force is applied. Examples of such fluids are water or gasoline.

For a Newtonian fluid, the ratio $\mu_a = \dfrac{\tau}{\dot{\gamma}}$, is constant at constant temperature and pressure, and is the viscosity.

1.6.2. Non-Newtonian fluids

We shall discuss only two types of non-Newtonian fluids which are most often encountered in drilling muds and cement slurries:

(a) Bingham fluids.
(b) Power-law fluids.

1.6.2.1. Bingham fluids

A. *Definition and typical curve*

In Bingham plastic fluids the shear stress also varies linearly with shear rate but, unlike Newtonian fluids, a minimum force must be applied to impart motion to them. This force, is known as the yield-point or yield-value.

Such fluids are characterized by two constants:

(a) Yield-point or yield-value τ_0, which corresponds to the smallest force required to set the fluid in motion.
(b) Plastic viscosity, μ_p, which is the ratio between the increment in the shear stress and the corresponding increment in the shear rate, i.e., the slope of the curve obtained by plotting τ as a function of $\dot{\gamma}$.

The theoretical equation of flow of such fluids is

$$\tau = \tau_0 + \mu_p \dot{\gamma} \tag{14}$$

In Cartesian coordinates, this is a straight line, as shown in the figure 9.

The experimental curves obtained with a rotational coaxial-cylinder viscometer are not strictly rectilinear, but rather as shown in the figure 10.

In fact, Eq. 14 is not applicable in the plug-flow zone.

B. *Determination of plastic viscosity μ_p and yield-point τ_0*

These determinations are carried out in a Fann viscometer according to *API* RP 13 B. The values to be determined are the shear stress $\tau_{1\,020}$ at the shear rate $\dot{\gamma}_2 = 1\,020$ s^{-1} (at 600 rpm), and the shear stress τ_{510} at the shear rate $\dot{\gamma}_1 = 510$ s^{-1} (at 300 rpm). These values are illustrated by the figure 11.

a. *Determination of the plastic viscosity μ_p.*

In SI units, we have
$$\mu_p = \frac{\tau_{1\,020} - \tau_{510}}{1\,020 - 510} \tag{15}$$

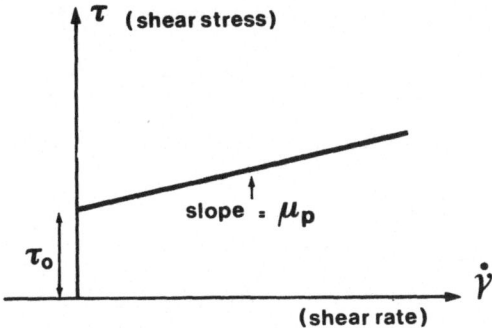

Fig. 9. — Bingham fluid theoretical flow curve.

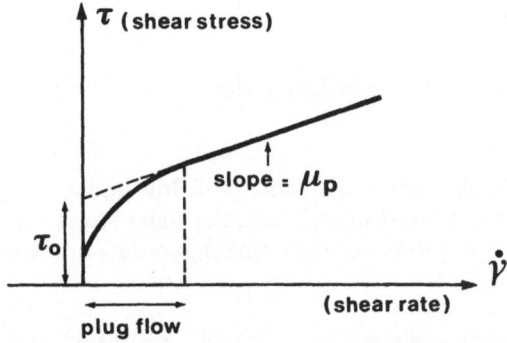

Fig. 10. — Bingham fluid experimental flow curve.

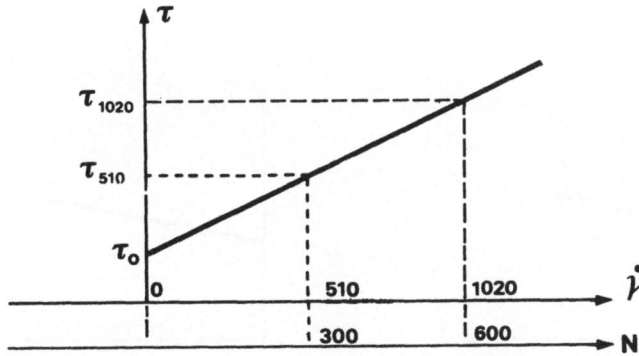

Fig. 11. — Determination of Bingham fluid rheological parameters.

If μ_p is expressed in centipoises, we have, in view of Eq. 5:

$$\mu_p = \frac{0.51(\theta_{600} - \theta_{300})}{1\,020 - 510}\,1\,000 = \theta_{600} - \theta_{300} \tag{16}$$

i.e., μ_p(cP) = Fann reading at 600 rpm − Fann reading at 300 rpm.

b. *Determination of the yield-point τ_0.*

It is apparent from the figure 11 that

$$\tau_0 = \tau_{1\,020} - 2(\tau_{1\,020} - \tau_{510}) \tag{17}$$

or, if τ_0 is expressed in lb/100 ft^2 and μ_a and μ_p in cP, in view of Eqs. 5 *bis*, 8 and 16:

$$\tau_0 = \theta_{600} - 2(\theta_{600} - \theta_{300}) \tag{18}$$
$$= \theta_{600} - 2\mu_p$$
$$= 2(\mu_a - \mu_p) \tag{18 bis}$$

1.6.2.2.   Pseudo-plastic or power-law fluids

A.   *Definition and typical curve*

Pseudo-plastic fluids, like Newtonian fluids, will flow under any applied stress, however small. But, as distinct from Newtonian fluids, the shear stress is not proportional to the shear rate, but to its nth power; hence the name power-law fluids. The equation of flow is

$$\tau = K\dot{\gamma}^n \tag{19}$$

where K is the consistency index in Pa . s^n or in lb . s^n/100 ft^2, and n is the dimensionless flow behavior index, which is unity or smaller than unity.

If $n = 1$, the equation becomes identical with the equation of flow of a Newtonian fluid having the viscosity K.

The following graphs shown in Figure 12 are flow curves of power-law fluid in Cartesian and logarithmic coordinates respectively.

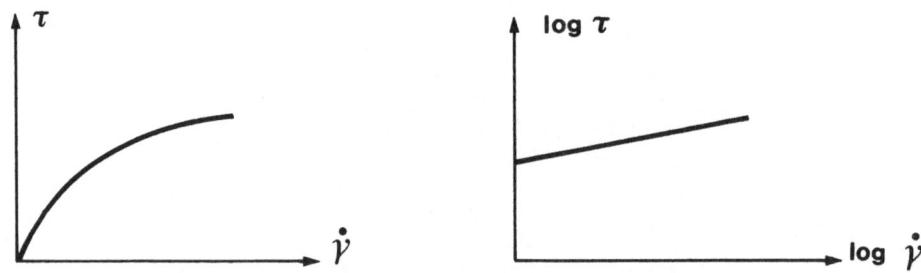

Fig. 12. − Power-law fluid flow curve.

B. *Fluid consistency index and flow behavior index*

In **logarithmic coordinates,** the flow curve is a straight line, the equation of which is

$$y = \log K + nx \tag{20}$$

where

$y = \log \tau$ and
$x = \log \dot{\gamma}$.

Thus, the flow behavior index n represents the slope of this line, while the fluid consistency index K is given by the intersection of the flow curve with the axis at $\dot{\gamma} = 1$:

$$n = \frac{\log \tau - \log \tau'}{\log \dot{\gamma} - \log \dot{\gamma}'} = \frac{\log \tau/\tau'}{\log \dot{\gamma}/\dot{\gamma}'} \tag{21}$$

**C. *Determination of flow behavior index n
and consistency index K in a Fann viscometer***

The determinations made in a six-speed Fann viscometer (or, if this instrument is not available, in a two-speed Fann viscometer, using also g_0, which is considered to represent a determination at 3 rpm) are plotted, as a rheogram, on log-log paper, shear rates (in s^{-1}) being plotted on the abscissa, shear stresses (in lb/100 ft^2) on the ordinate (Fig. 13).

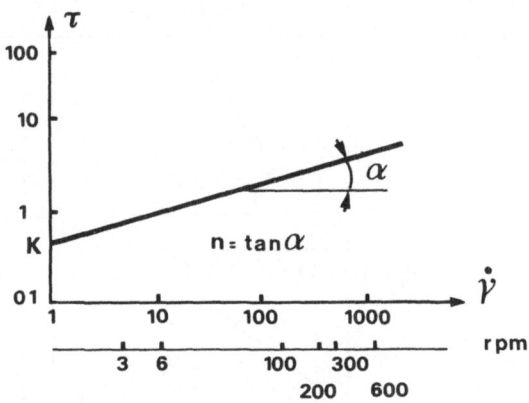

Fig. 13. — Determination of power-law fluid rheological parameters.

a. *Determination of n.*

We have seen that

$$n = \frac{\log \tau/\tau'}{\log \dot{\gamma}/\dot{\gamma}'} \quad \text{(dimensionless number)}$$

If $\dot{\gamma} = 2\dot{\gamma}'$, we have

$$n = \frac{\log \tau/\tau'}{\log 2} = \frac{\log \theta/\theta'}{\log 2} = 3.32 \log \frac{\theta}{\theta'} \tag{22}$$

Example:

$\dot{\gamma}_2 = 1\,020\ \text{s}^{-1}$ (at 600 rpm)

$\dot{\gamma}_1 = 510\ \text{s}^{-1}$ (at 300 rpm)

$$n = 3.32 \log \frac{\text{Fann reading at 600 rpm}}{\text{Fann reading at 300 rpm}} \tag{23}$$

b. *Determination of K*

$$K = \frac{\tau}{\dot{\gamma}^n} \tag{24}$$

If $\dot{\gamma} = 1$, $K = \tau_1$.

If τ is given in lb/100 ft^2 and $\dot{\gamma}$ in s^{-1}, the unit of K will be lb . s^n/100 ft^2. If τ is given in pascal the unit of K will be Pa . s^n. It will be recalled that 1 lb$_{\text{force}}$/100 ft^2 = 0.478964 Pa.

1.6.2.3. Summary of shear rate-shear stress relations

TABLE IV

FLOW	RHEOLOGICAL EQUATION	FLOW CURVE IN CARTESIAN COORDINATES	FLOW CURVE IN LOGARITHMIC COORDINATES
Newtonian	$\tau = \mu\dot{\gamma}$	arc tg μ	45°
Bingham plastic	$\tau = \tau_0 + \mu_p\dot{\gamma}$	arc tg μ_p	
Pseudo-plastic power-law	$\tau = K\dot{\gamma}^n$		log K arc tg n log 1

1.7. SHEAR THINNING

The apparent viscosity of the non-Newtonian Bingham and power-law fluids decreases during laminar flow as the shear rate increases. This effect is known as "shear thinning".

It will be recalled that the apparent viscosity is, by definition, $\mu_a = \dfrac{\tau}{\dot{\gamma}}$.

For Newtonian fluids, $\mu_a = \mu = $ constant.

For Bingham fluids, $\mu_a = \dfrac{\tau_0}{\dot{\gamma}} + \mu_p$, i.e., μ_a decreases with increasing $\dot{\gamma}$.

For power-law fluids $\mu_a = K\dot{\gamma}^{n-1} = \dfrac{K}{\dot{\gamma}^{1-n}}$, i.e., μ_a decreases as $\dot{\gamma}$ increases, for $n < 1$.

The following graphs of the Figure 14 represent the variations of the apparent viscosity with $\dot{\gamma}$.

In the case of pseudo-plastic fluids ($n < 1$), the extent of variation of μ_a with the shear rate is the larger, the smaller the value of n, i.e., the more they differ from Newtonian fluids.

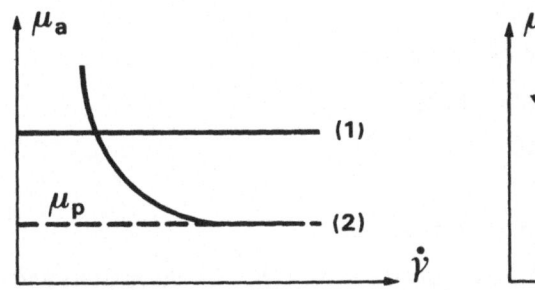 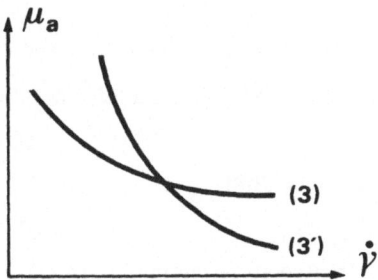

Fig. 14

1. Newtonian fluid.
2. Plastic Bingham fluid.
3, 3'. Pseudo-plastic power-law fluids; $n' < n$.

1.8. FLOW IN AN ANNULUS

The equations describing flow in pipes may be generalized to apply to other forms of ducts such as concentric annular spaces or parallel plates. In such cases, the tube diameter D must be replaced by an equivalent diameter φ_e, defined by:

$$\varphi_e = \delta \sqrt{8\psi(k)} \qquad (25)$$

where

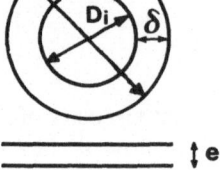

$$\delta = \frac{D_o - D_i}{2} \text{ for an annulus, or}$$

$\delta = e$ for two parallel plates,

$$k = \frac{D_i}{D_o}$$

and

$$\psi(k) = \frac{(1 + k^2) \ln k + (1 - k^2)}{2(1 - k)^2 \ln k} \tag{26}$$

in which, if

$$k \to 1 \qquad \psi(k) \to \frac{1}{3}$$

(in the limit, $k = 1$ becomes the case of two parallel plates. D_o and D_i are both infinitely large, so that $\dfrac{D_o}{D_i} = 1$, but $D_o - D_i \neq 0$). If however

$$k \to 0 \qquad \psi(k) \to \frac{1}{2}$$

(in the limit, $k = 0$, we have the case of a cylindrical pipe.)

NOTE. Thus, for two parallel plates, we have

$$\varphi_e = 0.8165 . 2e$$

    Annuli such that

$$\varphi_e \simeq 0.8165(D_o - D_i)$$

may be treated as a system of parallel plates, and the expressions for shear stress and shear rate become

$$\tau = \frac{(D_o - D_i)\Delta P}{4L} \tag{27}$$

$$\dot{\gamma} = \frac{12V}{D_o - D_i} \frac{(2n + 1)}{3n} \tag{28}$$

(Compare these expressions to the corresponding ones for a cylindrical tube (paragr. 1.4.2.7)).

2

Application to Drilling Fluids and Cement Slurries

2.1. INTRODUCTION

In this Chapter we shall present the general principles governing the calculation of the rheological parameters of drilling muds and cement slurries. All equations are given in a self-consistent system of units (SI).

The application of the results to laboratory practice, or in the field, will be discussed in Chapter 3.

The determination of the rheological characteristics of drilling fluids and cement slurries involves certain difficulties, for the following reasons :

A. *The properties of the fluids*

The fluids are non-Newtonian. They do not strictly conform to the limiting rheological laws which were discussed in Chapter 1, but display an intermediate type of behavior as illustrated in Figure 15.

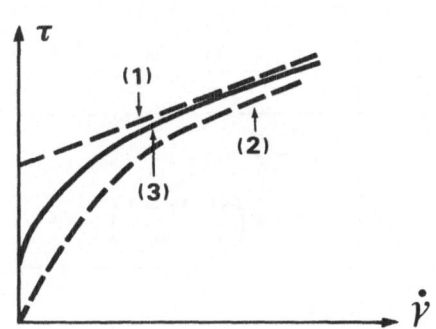 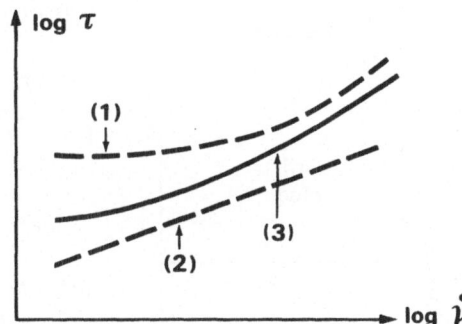

Fig. 15.
1. Bingham fluid.
2. Power-law fluid.
3. Drilling mud. Cement slurry.

The rheological characteristics are variable due to temperature and pressure variations in the well.

B. The geometry of the circulating system

This geometry is variable (irregular shape of the walls, eccentricity, etc.).
Annular flow is treated by approximate methods.

C. The complexity of the movements

In addition to the longitudinal flow:

(a) During the circulation of the mud, we have the movement imparted by rotation of drill string with resulting helical flow of the mud.
(b) During cementing, which may be accompanied by displacement of casing (reciprocating motion, rotary motion), an additionnal component of motion may be imparted.

A rigorous solution would be complex and cannot be justified in view of the approximate nature of the data and the level of accuracy needed.

TABLE V

EQUATIONS FOR EQUIVALENT VISCOSITY OF NON-NEWTONIAN FLUIDS

FLOW / MODEL		BINGHAM	POWER-LAW
Laminar	Drill pipe	$\mu_e = \dfrac{\tau}{8V/D}$ $= \mu_p + \dfrac{\tau_0}{8V/D}$	$\mu_e = \dfrac{\tau}{\dot{\gamma}}$ $= K\left(\dfrac{8V}{D}\dfrac{3n+1}{4n}\right)^{n-1}$
Laminar	Annular	$\mu_e = \dfrac{\tau}{12V/(D_o - D_i)}$ $= \mu_p + \dfrac{\tau_0}{12V/(D_o - D_i)}$	$\mu_e = \dfrac{\tau}{\dot{\gamma}}$ $= K\left(\dfrac{12V}{D_o - D_i}\dfrac{2n+1}{3n}\right)^{n-1}$
Turbulent	Drill pipe	$\mu_e = \mu_p$ (neglecting the effect of τ_0)	$\mu_e = \dfrac{\tau}{\dot{\gamma}}$ $= K\left(\dfrac{8V}{D}\dfrac{3n+1}{4n}\right)^{n-1}$
Turbulent	Annular		$\mu_e = \dfrac{\tau}{\dot{\gamma}}$ $= K\left(\dfrac{12V}{D_o - D_i}\dfrac{2n+1}{3n}\right)^{n-1}$

Accordingly, certain simplifications must be made; different simplications have been adopted by different authors. The ones adopted in this Chapter, which are also often encountered in drilling and cementing practice literature, are highly plausible; they may be summarized as follows:

(a) Irregularities in hole shape, motion of drill pipe and casing during the circulation, effects of temperature and pressure on the flow are neglected.

(b) The flow in the annulus is treated as taking place between two infinitely long parallel plates ([1]).

(c) The fluid is considered to be either of the Bingham or of the power-law type. The choice between the two is made according to Section 2.2 below.

(d)   At each shear rate, the fluid has a characteristic equivalent viscosity μ_e, as defined in Table V.

2.2.   SELECTION OF MODEL

As stated in the preceding paragraph, the rheological model chosen will be:

(a) Closest to the Bingham model, or
(b) Closest to the power-law model.

In most cases it is impossible to make this choice on the strength of the flow curve alone, whether plotted in Cartesian or in logarithmic coordinates. Certain calculations must be made, but this is now done easily and rapidly with the aid of programmable mini-computers.

The Fann viscometer yields most often only 4 experimental points because the readings at 3 and 6 rpm cannot be used if the fluid has a yield point (plug flow) or if it is thixotropic.

The four experimental points may be used for linear regression, in Cartesian and in logarithmic coordinates in succession. The model with correlation coefficient closest to unity, will be chosen for subsequent calculations. This correlation coefficient is given by the equation:

$$r = \frac{P\Sigma xy - \Sigma x\Sigma y}{\sqrt{[P\Sigma x^2 - (\Sigma x)^2][P\Sigma y^2 - (\Sigma y)^2]}}$$

where
  P is the number of Fann readings made,
  x is the shear rate or its logarithm ($\dot{y}$ or log $\dot{y}$), and
  y is the shear stress or its logarithm (τ or log τ).

NOTE.   In theory, this method is valid only if the difference between the two correlation coefficients is significant. A more rigorous method has recently been proposed in which the parameters a, b and c of the model equation $\theta = aN^b + c$, are found by an iterative program from the readings (θ) at 100, 200, 300 and 600 rpm [1].

([1]) Calculations show that for most annuli in wells, the equivalent diameter as defined in Section 1.8 is:
$$\varphi_e = 0.8165(D_o - D_i)$$

The values found for the parameters b and c are used as criteria in selecting the model, as follows:

(a) If $b \neq 1$, the Bingham model is selected.
(b) If $b < 1$ and $c \neq 0$, the power-law model is selected.

In the former case, the characteristics to be determined are the plastic viscosity μ_p and the corresponding yield-point τ_0. In the latter, the power-law index n and the corresponding consistency index K are determined.

2.3. DETERMINATION OF THE TYPE OF FLOW

The type of flow of a fluid is characterized by the Reynolds number Re, which is then compared to a **critical value** Re_c.

2.3.1. General expression for the Reynolds number and its critical value

Table VI shows the expressions for the Reynolds number and its generally accepted critical values for fluids of various rheological models circulating in pipes or in annuli.

NOTE. For the special case of a power-law fluid with $n = 1$, the equations become identical with those for a Newtonian fluid.

At the termination of the laminar flow, and before the turbulent flow begins, there is a transition zone, in which the flow is irregular, and the pressures in the fluid are highly

TABLE VI
EQUATIONS FOR REYNOLDS NUMBERS AND THEIR CRITICAL VALUES

MODEL	INTERIOR OF PIPES AND CASINGS	ANNULUS
Newton	$Re = \dfrac{VD\rho}{\mu}$ $Re_c = 2\,100$	$Re = \dfrac{0.8165(D_o - D_i)V\rho}{\mu}$ $Re_c = 2\,100$
Bingham	$Re = \dfrac{VD\rho}{\mu_e}$ $Re_c = 2\,100$	$Re = \dfrac{0.8165(D_o - D_i)V\rho}{\mu_e}$ $Re_c = 2\,100$
Power-law	$Re = \dfrac{VD\rho}{\mu_e}\dfrac{4n}{3n+1}$ $Re_c = 3\,470 - 1\,370n$	$Re = \dfrac{0.8165(D_o - D_i)V\rho}{\mu_e}\dfrac{3n}{2n+1}$ $Re_c = 3\,470 - 1\,370n$

unstable. Since pressure losses increase rapidly in turbulent flow, turbulent-flow equations should be used (for the sake of safety) as soon as laminar flow has ceased.

The **critical velocity,** V_c, is the velocity at the critical Reynolds number. If $V \leqslant V_c$, the flow is of the laminar type.

2.3.2.   Equations for Reynolds number and critical velocity as a function of the rheological parameters

2.3.2.1.   Newtonian fluids

$$Re = \frac{VD\rho}{\mu}$$

$$V_c = \frac{2\,100\mu}{D\rho}$$

2.3.2.2.   Bingham fluids

A.   *In drill pipes*

We have

$$Re = \frac{VD\rho}{\mu_e}$$

Substituting the expression for μ_e from Table V:

$$Re = \frac{VD\rho}{\dfrac{\tau_0 D}{8V} + \mu_p} = \frac{8V^2 D\rho}{\tau_0 D + 8V\mu_p} \;.$$

$$V_c = \frac{2\,100\mu_e}{D\rho} = 2\,100\,\frac{\mu_p + \dfrac{\tau_0}{8V_c/D}}{D\rho}$$

Hence

$$8V_c^2 D\rho - 16\,800V_c\mu_p - 2\,100\tau_0 D = 0$$

with V_c as the positive root of this second power equation.

$$V_c = \frac{16\,800\mu_p + \sqrt{(16\,800\mu_p)^2 + (4\,.\,8D\rho\,.\,2\,100\tau_0 D)}}{16D\rho}$$

$$V_c = \frac{16\,800\mu_p + \sqrt{(16\,800\mu_p)^2 + (32\,.\,2\,100\tau_0 D^2\rho)}}{16D\rho}$$

B.   *In annuli*

In the same way as before, we obtain:

$$Re = \frac{0.8165(D_o - D_i)V\rho}{\dfrac{\tau_0(D_o - D_i)}{12V} + \mu_p}$$

$$\text{Re} = \frac{9.8(D_o - D_i)V^2\rho}{\tau_0(D_o - D_i) + 12V\mu_p}$$

$$V_c = \frac{2\,100\mu_e}{0.8165(D_o - D_i)\rho} = 2\,572\,\frac{\mu_p + \dfrac{\tau_0}{12V_c/(D_o - D_i)}}{(D_o - D_i)\rho}$$

$$V_c = \frac{30\,864\mu_p + \sqrt{(30\,864\mu_p)^2 + 48 \cdot 2\,572\tau_0(D_o - D_i)^2\rho}}{24(D_o - D_i)\rho}$$

2.3.2.3.   Power-law fluids

A.   *In drill pipes*

The equations for Re and μ_e yield:

$$\text{Re} = \frac{VD\rho}{\mu_e}\,\frac{4n}{3n + 1} = \frac{VD\rho}{K\left(\dfrac{8V}{D}\,\dfrac{3n + 1}{4n}\right)^{n-1}}\,\frac{4n}{3n + 1}$$

$$\text{Re} = \frac{V^{2-n}D^n\rho}{K \cdot 8^{n-1}\left(\dfrac{3n + 1}{4n}\right)^n}$$

If Re = Re$_c$ = 3 470 − 1 370n, we obtain from this equation:

$$V_c = \left[\frac{(3\,470 - 1\,370n)K \cdot 8^{n-1}\left(\dfrac{3n + 1}{4n}\right)^n}{D^n\rho}\right]^{\frac{1}{2-n}}$$

B.   *In annuli*

We have, in this case:

$$\text{Re} = \frac{0.8165V(D_o - D_i)\rho}{\mu_e}\,\frac{3n}{2n + 1}$$

$$\text{Re} = \frac{0.8165V(D_o - D_i)\rho}{K\left(12\,\dfrac{V}{D_o - D_i}\,\dfrac{2n + 1}{3n}\right)^{n-1}}\,\frac{3n}{2n + 1}$$

$$\text{Re} = \frac{0.8165V^{2-n}(D_o - D_i)^n\rho}{K \cdot 12^{n-1}\left(\dfrac{2n + 1}{3n}\right)^n}$$

$$V_c = \left[\frac{(3\,470 - 1\,370n)K \cdot 12^{n-1}\left(\dfrac{2n + 1}{3n}\right)^n}{0.8165(D_o - D_i)^n\rho}\right]^{\frac{1}{2-n}}$$

2.4. CALCULATION OF PRESSURE LOSSES

2.4.1. Equations

The general equation for the pressure loss ΔP between two points separated by a distance L is:

$$\Delta P = \frac{2fL\rho V^2}{D} \qquad \text{for a cylindrical tube}$$

$$\Delta P = \frac{2fL\rho V^2}{\varphi_e} \qquad \text{for an annulus}$$

where f is the coefficient of head-pressure loss.

TABLE VII

EQUATIONS FOR PRESSURE LOSSES

		INSIDE PIPES AND CASINGS	ANNULUS
Laminar flow	Newton and Bingham fluids (Re < 2 100)	$f = \dfrac{16}{\text{Re}}$ $\Delta P = \dfrac{32L\mu_e V}{D^2}$	$f = \dfrac{16}{\text{Re}}$ $\Delta P = \dfrac{48L\mu_e V}{(D_o - D_i)^2}$
	Power-law fluids (Re < 3 470 − 1 370n)	$\Delta P = \dfrac{32L\mu_e V}{D^2}\dfrac{3n+1}{4n}$	$\Delta P = \dfrac{48L\mu_e V}{(D_o - D_i)^2}\dfrac{2n+1}{3n}$
Trubulent flow	Newton and Bingham fluids (Re ⩾ 2 100)	$f = \dfrac{0.05}{\text{Re}^{0.2}}$ $\Delta P = \dfrac{0.1L\rho^{0.8}V^{1.8}\mu_e^{0.2}}{D^{1.2}}$	$f = \dfrac{0.05}{\text{Re}^{0.2}}$ $\Delta P = \dfrac{0.127L\rho^{0.8}V^{1.8}\mu_e^{0.2}}{(D_o - D_i)^{1.2}}$
	Power-law fluids (Re ⩾ 3 470 − 1 370n)	$f = \dfrac{c}{\text{Re}^b}$ where $c = \dfrac{\log n + 2.5}{50}$ $b = \dfrac{1.4 - \log n}{7}$ $\Delta P = \left(\dfrac{2cL\rho^{1-b}V^{2-b}}{D^{1+b}}\right) \cdot$ $\cdot \left(\mu_e^b \left(\dfrac{3n+1}{4n}\right)^b\right)$	$f = \dfrac{c}{\text{Re}^b}$ where $c = \dfrac{\log n + 2.5}{50}$ $b = \dfrac{1.4 - \log n}{7}$ $\Delta P = \left(\dfrac{2cL\rho^{1-b}V^{2-b}}{[0.8165(D_o - D_i)]^{1+b}}\right) \cdot$ $\cdot \left(\mu_e^b \left(\dfrac{2n+1}{3n}\right)^b\right)$

It varies:

(a) With the type of the fluid.
(b) With the type of flow.
(c) With Reynolds number.

Table VII lists the general equations for f and ΔP for different types of fluids and flows. Expressions for the equivalent viscosity, μ_e, are given in Table V.

Equations for pressure losses applicable to the different rheological models are given in Chapter 3.

2.4.2. Calculation procedure

The calculation comprises the following steps:

(a) Determination of the flow type.
(b) Selection and utilization of the appropriate equation.

It can be represented by the following flowchart:

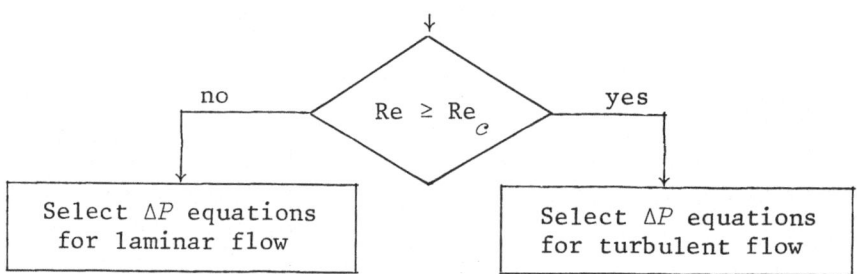

2.5. PRESSURE VARIATIONS DURING EQUIPMENT RUNNING OR PULLING

Moving a drill string (or a testing string or a casing) is accompanied by a displacement of the mud in the hole, leading to pressure changes.

Depending on the direction of this displacement, these changes may add to the pressure exerted by the mud, and produce a surge pressure, or, conversely, they may act in the opposite direction, and produce a swab effect. Such pressure variations may thus impair the stability of the hole wall even resulting in a blowout, or loss of circulation.

Pressure variations may be evaluated approximately by the method used in calculating pressure losses.

2.5.1.   Application of pressure to gelled mud

If the mud is left to stand, a gel will be formed due to its inherent thixotropy, and a definite pressure must then be applied to break up the gel and restore the fluidity of the mud. If the gel strength (long time gel strength) as determined in the Fann viscometer is g_t, the pressure which must be applied to restore the circulation of the mud may be found by the equation

$$P_r = \frac{4g_t L}{D_o - D_i}$$

where
g_t = long time gel strength $(g_t \geqslant g_{10})$,
L = section length,
D_o = hole diameter,
D_i = external diameter of string,
P_r = pressure.

Unfortunately, the gel strength determined at the surface may not be fully representative of the effective gel-strength in the hole, especially if the gelation of the mud takes place under the influence of heat.

The operator must insure that moving the mud from rest (moving the drill string or starting the pump) does not produce a dangerous overpressure. If at all possible, the gel should be broken up by rotating the string before mud circulation is begun.

2.5.2.   Pressure variations during pipe movement

The displacement of the mud produced by moving the drill string results in pressure variations. If the displacement velocities over the different sections of the annulus are known, the general equations for pressure loss during circulation are used.

2.5.2.1.   Theoretical displacement rate of the mud

The displacement velocity will depend on the speed of pipe movement and on the geometry involved (the mud may move inside an open string, but not inside a closed string). Assuming that the flow rate of the mud is equal to the volume of the string displaced per unit time, we find:

A.   Closed string

$$V_{th} = \frac{V_p D_i^2}{D_o^2 - D_i^2} + \frac{V_p}{2}$$

$$V_{th} = V_p \left[\frac{D_i^2}{D_o^2 - D_i^2} + \frac{1}{2} \right]$$

where
V_{th} = the theoretical velocity of the mud,
V_p = the speed of the string.

B. *Open string with complete filling*

This is the case of a string without a drill bit, of an open casing. A differentially filling casing must not be considered as an open string.

$$V_{th} = \frac{V_p(D_i^2 - D^2)}{D_o^2 - D_i^2 + D^2} + \frac{V_p}{2}$$

$$V_{th} = V_p\left[\frac{D_i^2 - D^2}{D_o^2 - D_i^2 + D^2} + \frac{1}{2}\right]$$

2.5.2.2   Equivalent displacement velocity of the mud

In actual practice, the displacement velocity of the mud to be used in the equations for pressure losses is slightly different from the theoretical velocity as a string in motion deforms the velocity profile (see Fig. 16).

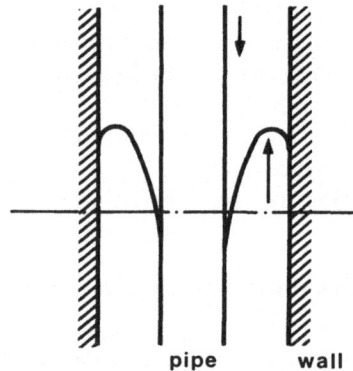

pipe   wall

Fig. 16. — Deformation of the velocity profile by the string motion.

The average equivalent velocities have been estimated by Burkhardt as:

$$V_e = V_p\left[\frac{D_i^2}{D_o^2 - D_i^2} + k\right]$$

and

$$V_e = V_p\left[\frac{D_i^2 - D^2}{D_o^2 - D_i^2 + D^2} + k\right]$$

where

V_e = the equivalent displacement velocity of mud,

k = is a function of the ratio $\dfrac{D_i}{D_o}$ and of the type of flow.

Burkhardt determined the value of k by the use of the Figure 17. A good general average value is $k = 0.45$.

Since the speed of the moving string is not constant, it seems that, on the average, the maximum fluid velocity is 1.5 times the average equivalent velocity

$$V_m = 1.5V_e$$

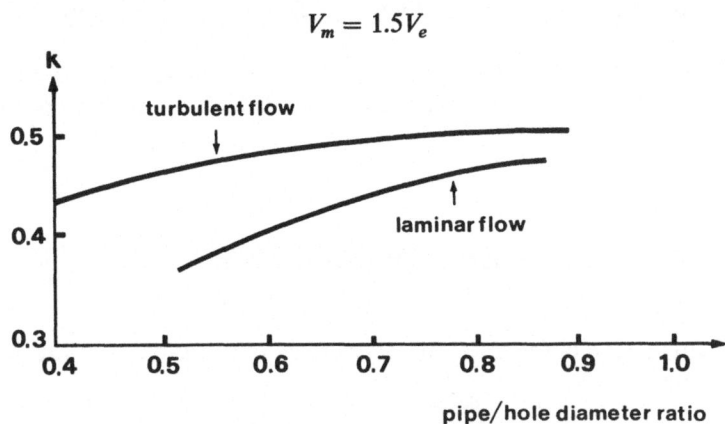

Fig. 17. — Values of k.

2.5.2.3. Calculation of surge pressures

Depending on whether we are dealing with a Bingham or with a power-law fluid, and on whether the flow corresponding to the equivalent velocity of the mud is laminar or turbulent, suitable equations for the calculation of pressure losses (Table VII) and of equivalent circulating density (Section 2.6) should be used.

Obviously, surge pressures are additive, just like pressure losses, and must be determined in each section of the annulus. However, it is not always easy to select the values of the rheological parameters to be used in these equations, because their values, as determined at the surface, may not be representative of down hole conditions (e.g., in the case of non-Newtonian fluids, whose viscosity is a function of the shear rate, temperature, and time).

2.5.3. Effect of inertia forces

A surge pressure is produced by the tendency of the mud column to oppose the variations in the displacement rate during the acceleration period immediately following the start of the circulation. In a similar manner, swab pressure is produced during the deceleration stage just before the movement of the mud is stopped.

According to Clark and Burkhardt, the pressure variation due to inertial effects is given by

$$\Delta P = \frac{L\rho D_i^2 a_p}{D_o^2 - D_i^2} \qquad \text{for a closed string}$$

$$\Delta P = \frac{L\rho(D_i^2 - D^2)a_p}{D_o^2 - D_i^2 + D^2} \qquad \text{for an open string}$$

In these equations a_p is the acceleration ($a_p > 0$) or the deceleration ($a_p < 0$). In most cases, this inertial effect is stronger than the gel-breakup pressure when the mud is set in motion and, if the accelerations are sudden, significantly larger pressure variations may result.

A different approach to the calculation of pressure variations during pipe movement is proposed by Lubinski [2]. These variations are considered as a transient rather than a permanent effect.

2.6. THE CONCEPT OF EQUIVALENT DENSITY

During the circulation of the mud, the pressure P' in the annulus at the depth L is the sum of two terms:

(a) The hydrostatic pressure exerted by the mud at this depth (P_h).
(b) Pressure losses due to the movement of the fluid between this depth and the surface (ΔP_a):

$$P' = P_h + \Delta P_a$$

2.6.1. Hydrostatic pressure (P_h)

It is given by the equation $$P_h = \rho_d g L$$

where g is the acceleration due to gravity.

The term ρ_d (fluid density) must allow for the presence of cuttings carried by the mud. We shall adopt the following notations:

q_d = volume of cuttings drilled per unit time,
Av = drilling rate,
D_f = bit diameter,
M_d = weight of cuttings drilled per unit time, on the assumption that the density of the cuttings is 2 500 kg/m³,
Q = injection rate,
ρ = density of injected mud,
ρ_d = density of the mud containing the cuttings.

We then have
$$q_d = \frac{\pi D_f^2 Av}{4}$$

$$M_d = 2.5 q_d$$

$$\rho_d = \frac{\rho(Q - q_d) + M_d}{Q}$$

$$\rho_d = \rho + \frac{q_d(2.5 - \rho)}{Q}$$

2.6.2. Pressure loss in the annulus ΔP_a (or in a part of the annulus)

Such losses are calculated using the equations in Section 2.4 (Table VII).

2.6.3. Equivalent circulating density and equivalent circulating specific gravity

The dimensions of density are ML^{-3}.

Specific gravity is a dimensionless number.

The equivalent circulating density and the equivalent circulating specific gravity are, respectively, the density and the specific gravity of the mud which would exert this pressure under static conditions at the depth L.

The equivalent circulating density is given by

$$\rho' = \frac{P'}{gL} = \rho_d + \frac{\Delta P_a}{gL}$$

The equivalent specific gravity is a dimensionless decimal number with numerical value equal to that of the density in SI units divided by 1 000.

The equivalent circulating density, or equivalent circulating specific gravity may be calculated at the bottom of the hole or at any other depth. The difference between the equivalent circulating density (or equivalent circulating specific gravity) and the true density (or true specific gravity) of the mud should be made as small as possible, in order to minimize the pressure difference between the static and dynamic conditions in the hole.

2.7. CLEANING THE HOLE AND MECHANICAL STRENGTH OF THE HOLE WALLS

One of the functions of the drilling fluid is to keep the hole clean, i.e., to ensure a proper removal of the cuttings; however, this must not be accompanied by erosion of hole walls. These two problems — cleaning and erosion — will now be discussed.

2.7.1. Cleaning the hole

Two stages must be considered:

(a) The entry of the cuttings produced by the bit into the mudstream.
(b) The lift of the cuttings in the annulus to the surface, towards the separation units.

2.7.1.1. Entry of cuttings from the bottom of the hole into the mudstream

It is important to prevent the rock fragments broken by the bit from the hold down effect at the bottom of the hole. The separation of the cuttings from the bottom of the hole is assisted:

(a) By a low differential pressure across the bottom of the hole.

(b) By a large fluid volume ("filtrate") and a low interfacial tension, to insure a better penetration of the liquid between individual chips and between the chips and the bottom of the hole.

(c) By a suitable design of bit nozzles, in order to impart to the mud sufficient energy to scavenge the cuttings.

2.7.1.2. Lift of cuttings in the annulus

The problem is complex.

A. *Qualitative considerations*

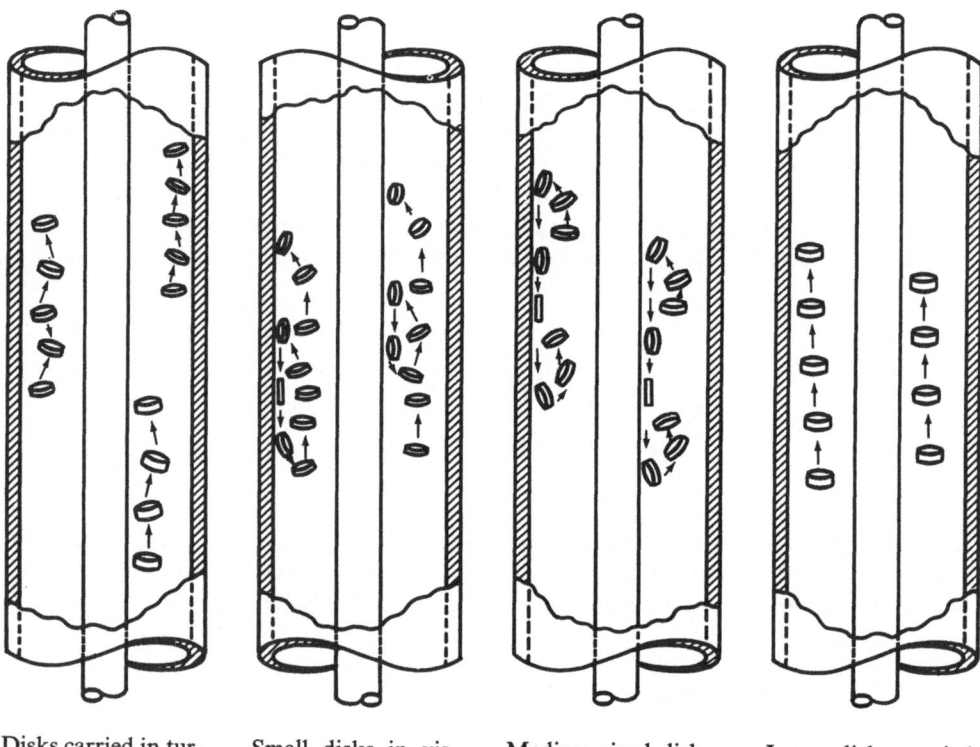

Disks carried in turbulent flow (pipe at rest)

Small disks in viscous flow (pipe at rest)

Medium-sized disks carried in viscous flow (pipe at rest)

Large disks carried in viscous flow (pipe at rest)

Fig. 18. — Lift of cuttings in drilling mud. After [3].

The cuttings in the annulus are acted upon by:
(a) The helicoidal ascending mudstream.
(b) The force of gravity.
(c) The centrifugal force.
(d) Unless the cuttings are spherical, they are also subject to a torque effect, which depends on the size and shape of the cuttings, their speed, profile and position in the annulus, as is schematically represented in Figure 18.

B. *Quantitative considerations*

a. *Computation of the uplift velocity of a cutting V_r*

Due to the complexity of the problem, only an approximate solution is possible. V_r is calculated by the equation

$$V_r = V_b - V_s$$

where
V_b = average uplift velocity of the mud,
V_s = terminal settling velocity of the cuttings (or slip velocity),
V_r = uplift velocity of the cuttings.

b. *Computation of the slip velocity V_s*

The equation to be used in calculating the slip velocity must involve the type of fluid flow pattern around the particle, that is:
(a) The mode of slip of the particle.
(b) The type of flow of the fluid.

In fact, no satisfactory solution to the problem has yet been found. Studies reported in the literature merely deal with partial aspects.

c. *Slip velocity of a solid particle in a fluid at rest*

1. The volume of the fluid is assumed to be infinitely large.

A synopsis of the methods and equations (in consistent units) to be used for various modes of particle slip is given in Table VIII. The meaning of the symbols utilized in the table is given below:
v = particle volume,
s = projected area (i.e., surface area of the projection of the particle onto a plane perpendicular to its direction of motion at the velocity V_s), (see Fig. 19).
ρ_s = particle density,
ρ_l = liquid density,
μ = liquid viscosity,
g = acceleration due to gravity,
C_r = resistance coefficient.
If the particle is not spherical, its equivalent diameter, i.e., the diameter of a sphere having an equal volume v is determined by

$$\varphi_p = \sqrt[3]{\frac{6v}{\pi}}$$

TABLEAU VIII

SLIP VELOCITY OF A PARTICLE IN A FLUID AT REST

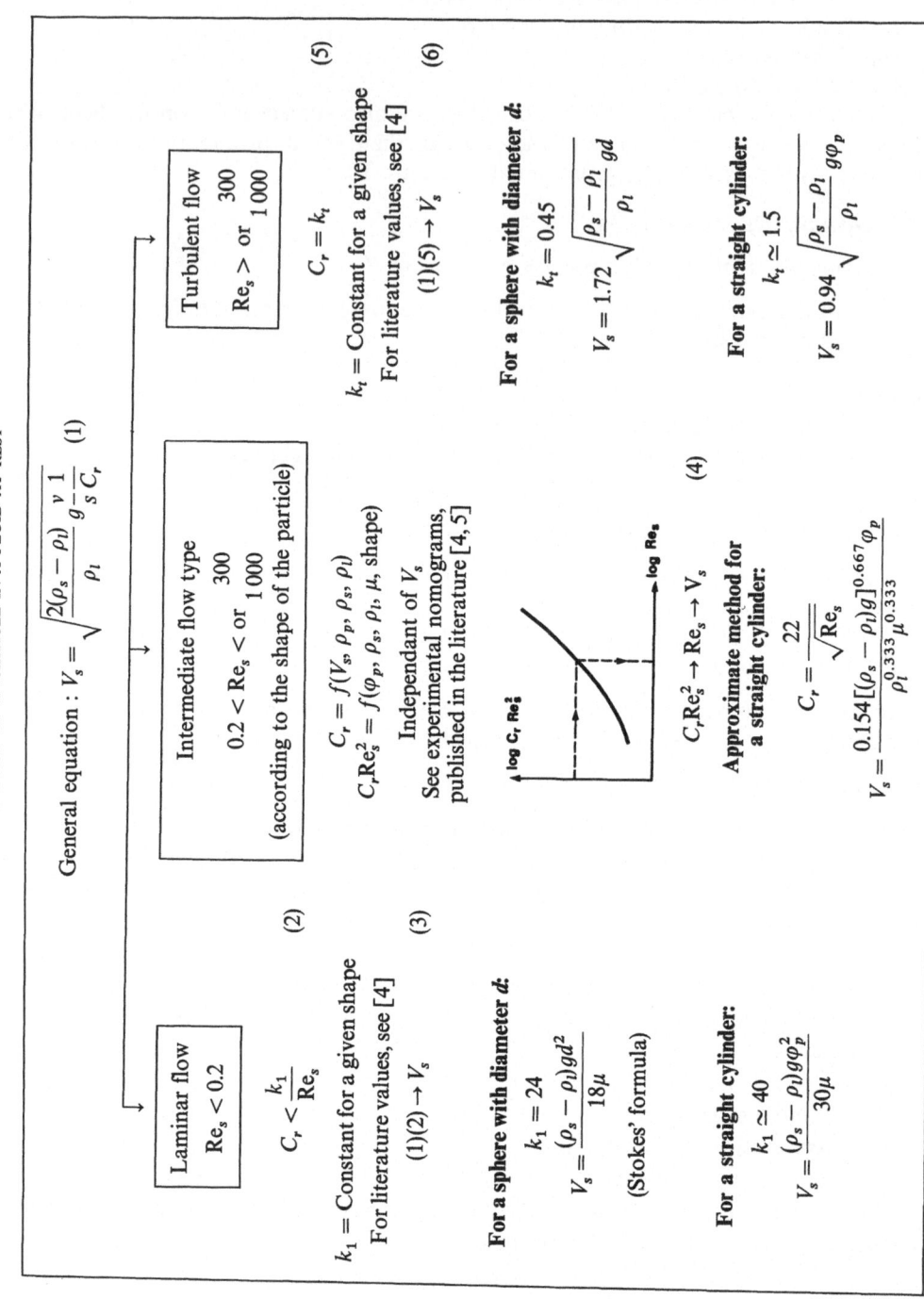

General equation : $V_s = \sqrt{\dfrac{2(\rho_s - \rho_l)}{\rho_l}\, \dfrac{v}{g}\, \dfrac{1}{s}\, C_r}$ (1)

Laminar flow
$Re_s < 0.2$

$C_r < \dfrac{k_1}{Re_s}$ (2)

$k_1 =$ Constant for a given shape
For literature values, see [4]

$(1)(2) \rightarrow V_s$

For a sphere with diameter d:

$k_1 = 24$

$V_s = \dfrac{(\rho_s - \rho_l)gd^2}{18\mu}$

(Stokes' formula)

For a straight cylinder:

$k_1 \simeq 40$

$V_s = \dfrac{(\rho_s - \rho_l)g\varphi_p^2}{30\mu}$

Intermediate flow type

$0.2 < Re_s < \begin{smallmatrix}300\\or\\1\,000\end{smallmatrix}$

(according to the shape of the particle)

$C_r = f(V_s, \rho_p, \rho_s, \rho_l)$
$C_r Re_s^2 = f(\varphi_p, \rho_s, \rho_l, \mu, \text{shape})$

Independant of V_s
See experimental nomograms,
published in the literature [4, 5]

$C_r Re_s^2 \rightarrow Re_s \rightarrow V_s$ (4)

**Approximate method for
a straight cylinder:**

$C_r = \dfrac{22}{\sqrt{Re_s}}$

$V_s = \dfrac{0.154[(\rho_s - \rho_l)g]^{0.667}\varphi_p}{\rho_l^{0.333}\mu^{0.333}}$

Turbulent flow
$Re_s > \begin{smallmatrix}300\\or\\1\,000\end{smallmatrix}$

$C_r = k_t$ (5)

$k_t =$ Constant for a given shape
For literature values, see [4]

$(1)(5) \rightarrow V_s$ (6)

For a sphere with diameter d:

$k_t = 0.45$

$V_s = 1.72 \sqrt{\dfrac{\rho_s - \rho_l}{\rho_l}\, gd}$

For a straight cylinder:

$k_t \simeq 1.5$

$V_s = 0.94 \sqrt{\dfrac{\rho_s - \rho_l}{\rho_l}\, g\varphi_p}$

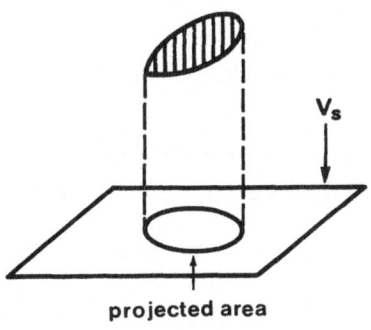

Fig. 19

The Reynolds number of the particle is

$$\text{Re}_s = \frac{V_s \varphi_p \rho_l}{\mu}$$

One of the difficulties involved in calculating the settling velocity in a non-Newtonian fluid is the selection of the viscosity value. This question will be discussed in the following paragraph.

2. Effect of the walls on the slip velocity.

Different equations, with allowance for the size of the particle, φ_p, relative to the size of the annulus, may be found in the literature.

For turbulent flow, we have

$$V_s(\text{corrected}) = \frac{V_s}{1 + (\varphi_p/\varphi_h)}$$

where φ_h is the hydraulic diameter of the annulus ($\varphi_h = D_o - D_i$).

For the intermediate type of flow [6] we have, for particles settling in a tube of diameter D:

$$V_s(\text{corrected}) = F_T V_s$$

where

$$F_T = \frac{D - 1.6\varphi_p}{D - \varphi_p}$$

However, the authors do not say whether this equation, applied to an annulus, should be

$$D \simeq \varphi_h = D_o - D_i$$

or

$$D \simeq \frac{\varphi_h}{2} = \frac{D_o - D_i}{2}$$

d. *Slip velocity in a circulating fluid*

According to different workers, the type of viscosity to be introduced into the above equations may be:

(a) The equivalent viscosity μ_c resulting from the circulation of the fluid [7, 8], or
(b) The equivalent viscosity μ_s resulting from the movement of the particle [6].

1. In the former case, μ_c may be calculated by two different methods:
(a) μ_c is equal to the equivalent viscosity μ_e as defined at the beginning of this Chapter.
(b) μ_c is equal to the "average" viscosity according to Walker [9] — if required, with allowance for pipe rotation. The average of the viscosities at 7 points is determined, with the segment midpoints subdividing the tickness of the annulus into seven equal parts.

2. In the latter case, the equivalent viscosity μ_s is derived from the shear stress and the shear rate resulting from the slip of the particle. Such a method has recently been proposed for the determination of the sedimentation rate of disks in the mud [6]. However, there are no generally valid definitions of the shear stress or shear rate, and the entire subject needs further study.

As nothing more definite is known on this problem, it would appear expedient, for practical purposes, to use the smaller of the two values μ_s and μ_c in the equations for V_s. Table IX recommends computing methods to be used for the various types of fluid flow and modes of slip of the cuttings.

TABLE IX

SLIP VELOCITY IN FLOWING FLUIDS

FLOW PATTERN	SLIP	$\dfrac{\mu_c}{\mu_s}$	VISCOSITY TYPE	EQUATION TO BE USED FOR V_s (TABLE VIII)
Laminar	Laminar	> 1 < 1	μ_s μ_c	(3)
	Intermediate	> 1 < 1	μ_s μ_c	(4)
	Turbulent		μ_s	(6)
Turbulent	Laminar Intermediate Turbulent			(6) (6) (6)

e. *Procedure for calculating V_s*

To calculate V_s, we must choose an equation fitting the mode of slip as characterized by the Reynolds number Re_s, which is itself a function of V_s.

In practice, the conditions of lift of the cuttings in the annulus are such that the laminar type of slip is **never** attained; the mode of settling is always intermediate or turbulent.

If the flow of the mud in the annulus is laminar, it is seen in Table IX that the equation to be used is Eq. 4 or 6, for intermediate or turbulent type slip, respectively.

The calculation consists of the following steps:

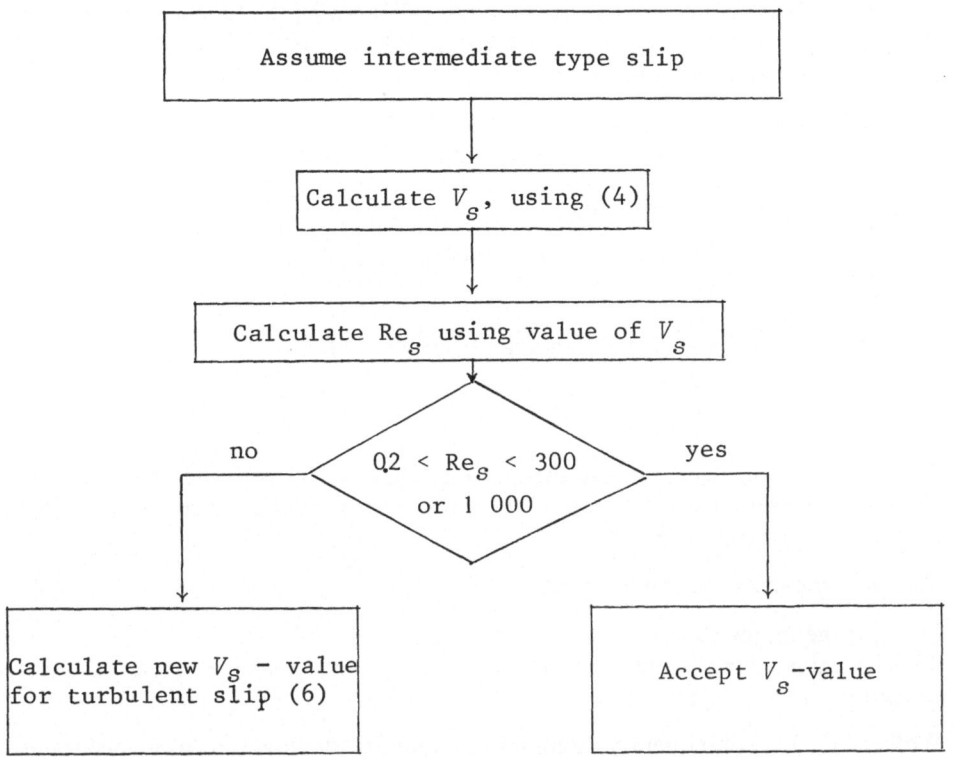

NOTE. In problems of sedimentation other than the settling of cuttings in the annulus during drilling which involve the fall of small particles in a viscous fluid (e.g., gravel-packing problems, hydraulic fracturing, etc.), the experimental value of Re_s may well be less than 0.2. In such cases, the slip velocity is calculated by Eq. 3 (Stokes' zone).

2.7.2. Mechanical strength of hole walls

2.7.2.1. Causes of erosion and weakness of walls

Erosion and weakness of the walls may have several origins:

(a) The shear of the drilling fluid against the hole wall during the circulation.
(b) Suction effect during pipe movement.
(c) Eccentricity of pipe with respect to the hole.
(d) Impact of the solid particles in the mud on the walls.
(e) Physical and chemical interaction between the mud and the exposed formation.
(f) Time of contact between the mud and the formation.

2.7.2.2. Effect of flow pattern of the mud

In the view of several workers, the flow pattern has a major effect on the stability of the hole walls. Walker [10] defined an erosion index (EI), which is a function of the flow type and of the shear stress τ_p at the hole wall, opposite the drill collars.

Since the type of flows is specified by Reynolds number, viz.:

(a) For laminar flow (Re < Re_c)

$$EI = k\tau_p$$

(b) For turbulent flow (Re > Re_c)

$$EI = (k\tau_p)^2$$

with τ_p in pascal, $k = 0.1$

τ_p is calculated with allowance for pipe rotation [11], as

$$\tau_p = \sqrt{\tau_{rz}^2 + \tau_{r\theta}^2}$$

where

τ_{rz} is the axial component of the shear stress, which varies with the pressure loss, and
$\tau_{r\theta}$ is the tangential component of the shear stress, which varies with the torque generated by pipe rotation.

Calculation shows that, most often:

(a) τ_{rz} is the major term.
(b) The value of τ_p is significantly less than the resistance to shear of the rock constituting the walls of the hole.

Walker [12, 13] subsequently attempted to describe the flow type more accurately by using the dimensionless function $Z(r)$ (cf. paragr. 2.7.2.3) instead of Reynolds number. In an annulus the value of Z varies with the distance r from the pipe wall.

If the maximum value of $Z(r)$ is less than 808, the flow is laminar; if it is higher, the flow is turbulent. According to this worker, the risk of erosion is enhanced if the value of Z near the hole wall attains 900.

La *Compagnie Française des Pétroles* (CFP) [14, 15] recently developed a practical method for minimizing the wall erosion in different wells situated in a given area, provided experimental results are already available for one or more other wells in the vicinity.

Pipe rotation is neglected. It is assumed that a wall will resist erosion of it meets the following condition: the Reynolds number opposite this wall must be lower than a certain statistical value, obtained during previous drillings conducted under similar conditions, i.e. as regards the nature of the ground, the drilling fluid, etc.

The flow rates, the rheological characteristics of the mud, and/or the diameters of the drill collars are adjusted with allowance for the period of time during which they remain opposite the zone under study. The Reynolds number is calculated with allowance for the equivalent viscosity of the fluid at the wall, by the equations given in Table V.

However, this method cannot be employed if mechanical erosion produced by percussion by the drill string is not negligeable.

2.7.2.3.   The Z-number

For non-Newtonian liquids, Reynolds number is only an approximate criterion of the type of flow.

Ryan and Johnson [14] proposed another, more rigorous, experimental criterion. They consider a dimensionless function $Z(r)$ which, in the case of longitudinal flow in a straight cylindrical pipe, is given by

$$Z(r) = \frac{-\rho V(r)(\mathrm{d}V/\mathrm{d}r)}{\mathrm{d}\tau/\mathrm{d}r}$$

$$Z(r) = -\frac{R}{\tau(R)}\rho V(r)\frac{\mathrm{d}V}{\mathrm{d}r}$$

where
  R = radius of pipe,
  r = distance from the axis of the pipe,
  V = longitudinal velocity,
  ρ = density of fluid,
  τ = shear stress.

$Z(r)$ varies with r, and its maximum value is the criterion of the type of flow. If the maximum $Z(r)$ value is below 808, the flow is laminar. If it is above 808, it is turbulent.

NOTE.   The value of 808 has been so chosen that, for a Newtonian fluid (with a parabolic velocity profile) the Reynolds number is 2 100. We therefore have

$$Z = 2\sqrt{\frac{1}{27}}\,\mathrm{Re}$$

The validity of this criterion has been repeatedly checked for the different cases, including Newtonian and non-Newtonian fluids flowing in ducts of various shapes (cylindrical, annular, parallel plates), but always parallel to the duct axis [17, 18, 19].

Walker [12, 13] subsequently recommended the application of tne criterion of Ryan and Johnson to the flow of drilling mud in an annulus, with allowance for pipe rotation (helicoidal motion).

In the case of a power-law fluid, $\dot{Z}$ is given by

$$Z = \frac{(D_o - D_i)^n V^{2-n}\rho}{K}(\psi)$$

where ψ is a function of n, of the geometry and of the angular velocity of the pipe.

However, studies by other workers, made at about the same time, failed to conclusively confirm the validity of the Ryan-Johnson criterion for helicoidal flow [20].

2.8. HYDRAULIC POWER

2.8.1. Definition and source

The surface pumping capacity is limited:

(a) By the power of the equipment employed, viz.:
    . By the power of the motors (which varies with their working conditions and with the temperature of the environment).
    . By the rated output of the pump.
(b) By the equipement efficiency.

If $\mathscr{P}_M$ is the motive power and η_T the transmission efficiency (which may vary between 0.65 and 0.90, depending on whether the installation is or is not provided with a torque converter), the mechanical power $\mathscr{P}_m$ at pump inlet is

$$\mathscr{P}_m = \eta_T \mathscr{P}_M \tag{1}$$

If the mechanical efficiency of the pump is η_m (about 0.80), its theoretical hydraulic power would be

$$\mathscr{P}_{h_{th}} = \eta_m \mathscr{P}_m$$

In practice, the hydraulic efficiency η_v is 0.90-0.95, and the maximum available hydraulic power is

$$\mathscr{P}_{h_{l \max}} = \eta_v \mathscr{P}_{h_{th}} = \eta_v \eta_m \eta_T \mathscr{P}_M \tag{2}$$

This hydraulic power is proportional to the product of the flow rate Q and the injection pressure P_i.

Q is the actual flow rate measured in the suction pit. P_i is equal to the sum of pressure losses in the entire mud circuit.

We therefore have

$$\mathscr{P}_h = Q \Delta P \tag{3}$$

The theoretical flow rate varies with the liner size, the stroke of the piston, the diameter of piston rods and the pumping rate.

The maximum service pressure is specified by the manufacturer of the pump for the given liner and the given nominal rate.

For reasons of economy, the hydraulic power actually utilized should not be more than 75% of the maximum available power. We thus have

$$\mathscr{P}_{h_l} = 0.75 \mathscr{P}_{h_{l \max}}$$

NOTE.   The maximum service pressure of the circulation system is the lower value between the maximum service pressure of the pump and the maximum service pressure of the surface connections.

2.8.2. The role of hydraulic conditions

If, in a given hydraulic circuit, the flow rate is increased, more power is expended in pressure losses, and the residual power available to the bit will decrease, and *vice versa*. It is clear, accordingly, that there is some optimum flow rate at which the maximum power can be transmitted to the bit.

The optimum working conditions of the mud at the bit are described in terms of two concepts:

(a) Hydraulic power.
(b) Hydraulic impact.

2.8.2.1. Hydraulic power at the bit

This magnitude is defined, as before, by the product:

$$\mathscr{P}_{h_e} = Q \Delta P_e \tag{4}$$

where ΔP_e is the pressure loss in the bit. The dimensions of this product are ML^2T^{-3}.

The hydraulic power $\mathscr{P}_{h_e}$ is the power developed by the mud on passing through the bit nozzles.

2.8.2.2. Hydraulic impact at the bit I_h

This is defined as follows

$$I_h = \rho Q V = \rho \frac{Q^2}{A} \tag{5}$$

where
V is the average velocity of the mud passing through the bit nozzles, and
A is the total cross-sectional area.

This product has the dimensions of a force, i.e., MLT^{-2}. The hydraulic impact I_h is the force exerted by the mud at the bottom of the hole.

Numerous field trials have demonstrated that, for a given level of hydraulic power at the bit $(\mathscr{P}_{h_e})_1$ or a given level of hydraulic impact $(I_h)_1$, the drilling rate Av is proportional to the weight acting on the bit, up to a limit value Av_1.

If the hydraulic level is increased to a value $(\mathscr{P}_{h_e})_2$ or $(I_h)_2$, the weight acting on the bit can be increased up to a new plateau, which will be attained at Av_2.

The plateaus thus defined are due to the imperfect cleaning of the cuttings from the bottomhole.

2.8.3. Optimization of the values of $\mathscr{P}_{h_e}$ and I_h

Because the hydraulic power available for pumping and the maximum service pressure of the surface installations for a given system are limited (pump, surface connections, pipe, drill collars), the values of $\mathscr{P}_{h_e}$ and I_h should be optimized by varying the rate of flow and the diameter of the nozzles, with due allowance for the pressure losses in the bit and outside it.

All "hydraulic" computations proposed in the following paragraphs apply to a **Bingham fluid in turbulent flow.**

2.8.3.1. Pressure losses. Equations and limitations

A. *Pressure losses in the bit*

These may be expressed as follows

$$\Delta P = k_e \rho V^2 \tag{6}$$

where k_e is a proportionality factor depending on the type of the bit.

B. *Pressure losses outside the bit*

For a given mud, these losses vary with the depth. In accordance with the equation given in Section 2.4, they may be written

$$\Delta P_c = \frac{2fL\rho V^2}{D} = k_c Q^{1.8} \tag{7}$$

and hence,

$$\log \Delta P_c = \log k_c + 1.8 \log Q$$

In logarithmic coordinates, the variation of ΔP_c with Q is represented by a set of parallel straight lines with the slope of 1.8, each line for a different depth.

C. *Overall pressure losses in the circuit*

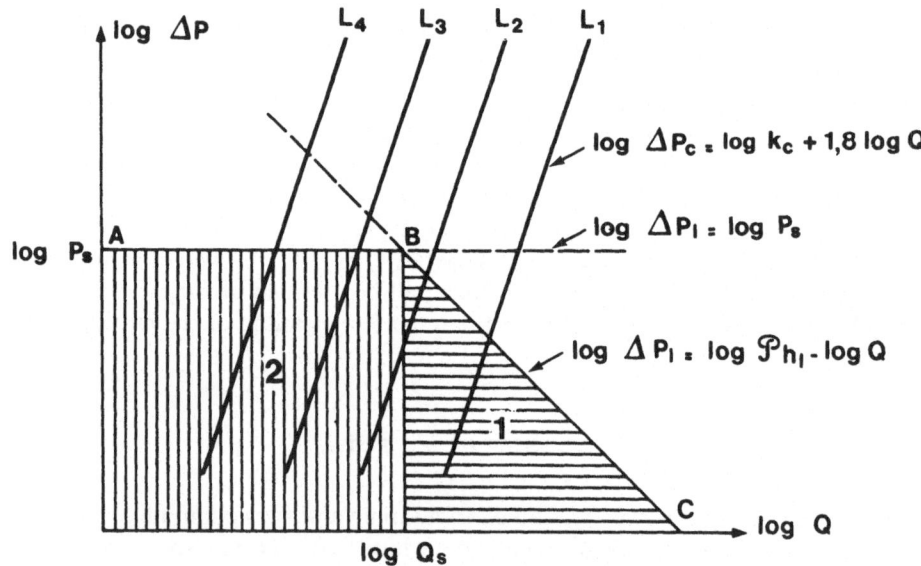

Fig. 20. — Plot of ΔP as a function of Q.

They are equal to the sum

$$\Delta P = \Delta P_e + \Delta P_c$$

They are limited upto the maximum value ΔP_l:

(a) On the one hand, by the service pressure P_s of the installation.

(b) On the other, by the maximum available hydraulic power, given by the relation

$$\Delta P_l = \frac{\mathscr{P}_{h_l}}{Q} \tag{8}$$

or

$$\log \Delta P_l = \log \mathscr{P}_{h_l} - \log Q.$$

In logarithmic coordinates, ΔP_l is represented as a function of Q ($\Delta P_l < P_s$) by a rectilinear segment with a slope of -1. The Figure 20 shows such a log-log plot for a constant geometry.

The maximum permissible total pressure-losses, ΔP_l, are represented by the linear segments AB and BC.

Two different zones can be distinguished in the Figure:

(a) Zone 1, in which the injection pressure is not limited by the service pressure of the surface installation.

(b) Zone 2, in which the injection pressure is limited by the service pressure of the surface installation.

2.8.3.2.   Optimum distribution of pressure losses

A.   *Criterion of maximum hydraulic pressure at the bit*

a.   *No limitation by surface pressure*

$$\mathscr{P}_h = \mathscr{P}_{h_c} + \mathscr{P}_{h_e}$$
$$\mathscr{P}_{h_e} = \mathscr{P}_h - \mathscr{P}_{h_c}$$
$$\mathscr{P}_{h_e} = Q\Delta P - Q\Delta P_c$$

and $\mathscr{P}_{h_e}$ will have its maximum value at the maximum value of $Q\Delta P - Q\Delta P_c$.

Whatever the value of Q, the value of $Q\Delta P$ will be largest if it is equal to the total available power.

$Q\Delta P_c$ will assume its minimum value when the value of Q is least, because ΔP_c varies in the same way as Q.

$\mathscr{P}_{h_e}$ will assume its maximum value when

$$\mathscr{P}_h = \mathscr{P}_{h_l}$$
$$Q = Q_{min}$$

where $Q_{min} \geqslant Q_s$

If this were not so, we would be dealing with case b., which will now be discussed.

b.   *Limitation by surface pressure*

$$\mathscr{P}_{h_e} = Q\Delta P - Q\Delta P_c$$

We have seen that $\Delta P_c = k_c Q^{1.8}$, so that

$$\mathscr{P}_{h_e} = Q\Delta P - k_c Q^{2.8} \tag{9}$$

For a given rate of flow, $\mathscr{P}_{h_e}$ will have the largest value for the largest ΔP, i.e., equal to ΔP_t (viz., $\Delta P = \Delta P_l = P_s$).

The maximum value of $\mathscr{P}_{h_e}$ as a function of Q is obtained when

$$\frac{d\mathscr{P}_{h_e}}{dQ} = 0$$

Differentiation of Eq. 9 yields:

$$\Delta P - 2.8 k_c Q^{1.8} = 0$$

and as

$$\Delta P_c = k_c Q^{1.8} \rightarrow \Delta P - 2.8\Delta P_c = 0$$

we obtain the following distribution pressure losses:

$$\boxed{\begin{array}{l} \Delta P_c = 0.35\Delta P \\ \Delta P_e = 0.65\Delta P \\ \Delta P \ = \Delta P_l = P_s \end{array}}$$

B.   Criterion of maximum hydraulic impact on the bit

From Eqs. 5 and 6:

$$I_h = \rho Q V$$

$$\Delta P_e = k\rho V^2 = k_e \frac{Q^2}{A}$$

Also

$$\Delta P_c = k_c Q^{1.8}$$

so that

$$\mathscr{P}_h = Q\Delta P_c + Q\Delta P_e$$

$$\mathscr{P}_h = k_c Q^{2.8} + k_e \frac{Q^3}{A^2}$$

$$\mathscr{P}_h = k_c Q^{2.8} + \frac{k_e I_h^2}{\rho^2 Q}$$

$$I_h^2 = \frac{\rho^2 \mathscr{P}_h Q}{k_e} - \frac{\rho^2 k_c Q^{3.8}}{k_e} \tag{10}$$

a.   No limitation by surface pressure

For a given rate of flow, and with $\mathscr{P}_h$ equal to the maximum available power, the hydraulic impact will have its maximum value when

$$\frac{dI_h}{dQ} = 0 = \frac{1}{2}\left[\frac{\rho^2 \mathscr{P}_h}{k_e} - 3.8\rho^2 \frac{k_c}{k_e} Q^{2.8}\right]$$

Hence $\mathscr{P}_h = 3.8k_cQ^{2.8}$ when $\Delta P_c = k_cQ^{1.8}$

$$Q\Delta P = 3.8Q\Delta P_c$$
$$\Delta P = 3.8\Delta P_c$$

The pressure losses are then distributed as follows:

$$\boxed{\begin{aligned} \Delta P_c &= 0.26\Delta P \\ \Delta P_e &= 0.74\Delta P \\ \Delta P &= \Delta P_l \end{aligned}}$$

b.　*Limitation by surface pressure*

If, in Eq. 10, $\mathscr{P}_h$ is replaced by $Q\Delta P$, we may write:

$$I_h^2 = \frac{\rho^2}{k_e} Q^2 \Delta P - \rho^2 \frac{k_c}{k_e} Q^{3.8}$$

The value of I_h is largest at the maximum ΔP-value, and is equal to the limit pressure $(\Delta P = \Delta P_l = P_s)$:

I_h, as a function of Q, will be largest when $\dfrac{dI_h}{dQ} = 0$, i.e.

$$\frac{dI_h}{dQ} = \frac{1}{2}\left[2\frac{\rho^2 Q\Delta P}{k_e} - 3.8\frac{k_c\rho^2}{k_e} Q^{2.8} \right] = 0$$

The factor between brackets becomes zero when

$$2\Delta P = 3.8k_cQ^{1.8}$$

As $\Delta P_c = k_cQ^{1.8}$

$$\Delta P = \frac{3.8}{2}\Delta P_c = 1.9\Delta P_c$$

The new distribution in the system becomes

$$\boxed{\begin{aligned} \Delta P_c &= 0.52\Delta P \\ \Delta P_e &= 0.48\Delta P \\ \Delta P &= \Delta P_l = P_s \end{aligned}}$$

2.8.3.3.　Note on the relationship between ΔP_c and Q and its consequence

In pratice, if the mud is not a Bingham fluid and/or the annular flow is not turbulent, the pressure losses outside the bit may be written

$$\Delta P_c = k_cQ^m$$

m is other than 1.8 and may be as low as 1.2.

Then, in case of limitation by surface pressure, the optimum distribution of pressure losses may be satisfactorily taken according to the following relations:

(a) Criterion of maximum hydraulic power at the bit

$$\Delta P_c = \frac{1}{1 + m} \Delta P$$

$$\Delta P_e = \frac{m}{1 + m} \Delta P$$

$$\Delta P = P_s$$

(b) Criterion of maximum hydraulic impact on the bit

$$\Delta P_c = \frac{2}{2 + m} \Delta P$$

$$\Delta P_e = \frac{m}{2 + m} \Delta P$$

$$\Delta P = P_s$$

2.9. MODIFICATIONS
OF RHEOLOGICAL PARAMETERS

2.9.1. Bingham fluids

Circulating Bingham fluids are described by two parameters: plastic viscosity and yield-point.

2.9.1.1. Plastic viscosity

This parameter is a function of:

(a) The concentration of solids.
(b) The size and shape of the solid particles.
(c) The viscosity of the liquid phase.

Plastic viscosity increases with increasing solid content or, for constant solid content, with increasing number of solid particles (fine particles), that is, with increasing specific particle surface. Conversely, it decreases with decreasing solid content or, for a given solid content, with decreasing number of solid particles (coarser particles), i.e., with decreasing specific surface (flocculation).

2.9.1.2. Yield-point

The yield-point results from cohesive forces between the particles, due to the electric charges on their surfaces. The magnitude of these forces will depend on:

(a) The type of the solids and their surface charges.
(b) The amount of the solids present.
(c) The ion concentration in the liquid phase.

High yield-points may be due to:

(a) Grinding of the solids by the bit, pipe, etc., with consequent increase in their specific surface area.
(b) Increase in solid content, with consequent decrease in inter-particle distance.
(c) Contamination by salt, gypsum, etc., which favours flocculation of the particles.
(d) Insufficient concentration of the thinning agent, the function of which is to neutralize the attractive forces.

If the factors responsible for the variability of these parameters are known, appropriate treatments can be applied.

Thus, plastic viscosity may be reduced (or its increase retarded) by de-sanding, de-silting, centrifuging, treatment in sophisticated vibrators or by dilution.

The yield-point will be reduced by the addition of substances neutralizing the electric charges, such as thinning agents, and by the addition of chemicals to precipitate the contaminants. If elimination of the contaminant is impossible (in the case of salt, for example), the yield-point may be reduced either by reducing the solid content (dilution) or by using a more suitable type of mud.

As a general rule, it would seem that plastic viscosity can be lowered by reducing the solid content, while the yield-point is more readily affected by chemical treatments.

2.9.2. Power-law fluids

Circulating power-law fluids are described by two parameters.

2.9.2.1. Consistency index K

The variation of K is due to the same influences as that of plastic viscosity, in the case of Bingham fluids.

2.9.2.2. Power exponent n

The value of the exponent n may be influenced by the same effects as the yield-point; however, effects which increase the yield-point, reduce the value of n.

The value of n may also be reduced by the introduction of specific additives such as:

(a) Bio-polymers.
(b) Pre-hydrated bentonite, wich has been flocculated by salts.

(c) "Super-bentonite" or chemically-treated bentonite.

(d) Certain carboxymethylcelluloses, guar gum, attapulgite, asbestos fibers, etc.

2.9.3. Gels

The initial and the 10 minute gel-strengths are an indication of the attractive forces operative in a static suspension.

If the difference between these two gel-strengths is large, the gel is known as "progressive". If there is practically no difference, the gel is known as "flat".

Gels are a measure of static attractive forces, while the yield-point is a measure of dynamic attractive forces. Thus, the yield-point and the gel-strength are increased or decreased by the same treatments.

2.9.4. Marsh funnel viscosity

This rough determination reflects all rheological parameters and other physical characteristics of the mud, but does not permit the identification of the principal parameters affecting the result of the measurement.

The numerical values obtained in this test cannot be used for any hydraulic calculations. They merely provide relative indications for muds of comparable types and physical properties.

REFERENCES

1   Ricard, G. — « Rhéologie des boues de forage et des ciments ». *Forages*, n° 76, p. 83-112, July-Sept. 1977.

2   Lubinski, A., Hsu, F. H., Nolte, K.G. — "Transient pressure surge due to pipe movement in an oil well". *Rev. Inst. Franç. du Pétrole*, XXXII, no 3, pp. 307-347, May-June 1977.

3   Williams, C. E., Bruce, G. H. — "Carrying capacity of drilling muds". *Pet. Trans AIME*, vol. 192, pp. 111-120, 1951.

4   Ludwig, J. — "Sinkversuche mit Festteilchen verscheidener Gestalt in Flüssigkeiten". *Chem. Ztg*, n° 22, pp. 774-777, 1955.

5   Rogers, W. F. — *Composition and properties of oil well drilling fluids*. Gulf Publishing Co, Houston, Chapter 6, pp. 308-316, 1953.

6   Walker, R. E., Mayes, T. M. — "Design for carrying capacity". *SPE*, n° 4975, Houston, Oct. 1974.

7   Udo Zeidler, H. — "An experimental analysis of the transport of drilled particles". *SPE*, n° 3064, Houston, Oct. 1970.

8   Sifferman, T. R. et coll. — "Drill cutting transport in full scale vertical annulus". *SPE*, n° 4514, Las Vegas, Oct. 1973.

9   Walker, R. E. — "Drilling fluid rheology". Notes from the *Kelco Drilling Fluid Seminars*, 1969.

10 Walker, R. E., Holman, H. E. — "Computer program predicts drilling fluid performance". *Oil and Gas J.*, pp. 80-90, March 29, 1971.

11 Walker, R. E., Al Rawi, O. — "Helical flow of bentonite slurries". *SPE*, n° 3108, Houston, Oct. 1970.

12 Walker, R. E., Korry, D. E. — "Field method of evaluating annular performance of drilling fluids". *SPE*, n° 4321, London, April 1973.

13 Walker, R. E. — "Mud Hydraulics".
5 — *Oil and Gas J*. 74-40, *pp*. 86-90, Oct. 4, 1976.
6 — *Oil and Gas J*. 74-42, *pp*. 82-88, Oct. 18, 1976.
7 — *Oil and Gas J*. 74-44, *pp*. 72-82, Nov. 1, 1976.

14 Parigot, P. — « Érosion de la paroi ». *Doc. CFP TEP/DP/FOR*, Feb. 1976.

15 *CFP*. — « Étude de l'érosion à la paroi. Approche statistique du problème par évaluation du nombre de Reynolds modifié ». *Doc. CFP TEP/DP/FOR*, March 1977.

16 Ryan, N. W., Johnson, M. M. — "Transition from laminar to turbulent flow in pipes". *A.I.Ch.E.J.*, 5-4, pp. 434-440, Déc. 1959.

17 Hanks, R. W. — "The laminar-turbulent transition for flow in pipes, concentric annuli and parallel plates". *A.I.Ch.E.J.*, 9-1, pp. 45-48, January 1963.

18 Hanks, R. W. — "The laminar-turbulent transition for fluids with a yield stress". *A.I.Ch.E.J.*, 9-3, pp. 306-309, March 1963.

19 Le Fur, B. et coll. — « Transition de l'écoulement de liquides non newtoniens dans des conduites circulaires ». *Comptes rendus 2ᵉ colloque ARTFP*, p. 113-130, Éditions Technip, 31 May-4 June 1965.

20 Latil, M., Martin, M. — « Écoulement des boues de forage dans les annulaires ». *Report IFP*, n° 20783, *CFP*, Dec. 1972.

3

Principal Methods of Evaluation

3.1. INTRODUCTION

When applying the results obtained in the preceding Chapters in practice, it is necessary:

(a) To use standard instruments for the determination of the rheological parameters of drilling fluids and cement slurries.
(b) To employ unit systems which are accepted in the field operations.

Table XII lists the parameters which are determined, and the different units in which they are expressed:

(a) SI units.
(b) American units.
(c) Practical units used by French organizations, named "other units".

Table XI lists the conversion factors.

3.2. SELECTION OF MODEL AND DETERMINATION OF RHEOLOGICAL PARAMETERS

3.2.1. Six-speed Fann viscometer

Linear regression (paragr 2.2) is used on the four useful experimental points (100, 200, 300 and 600 rpm).

After the model has been selected, the following rheological parameters are determined:

(a) τ_0 and μ_p for a Bingham fluid.
(b) n and K for a power-law fluid.

This determination may be made graphically or by the method of least squares.

3.2.1.1. Graphical determination of parameters for the Bingham model

The Fann viscometer readings are plotted in **Cartesian coordinates,** as in figure 21. μ_p is proportional to the slope of the best straight line drawn through the four points. τ_0 is proportional to the ordinate value at the origin of the straight line.

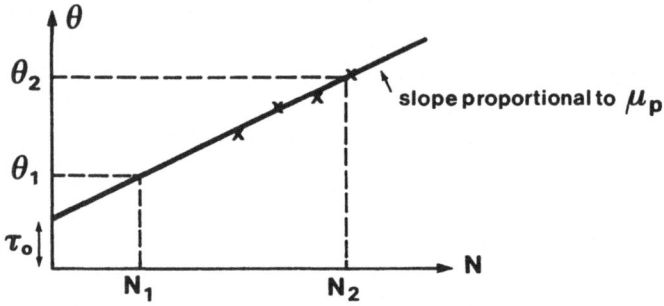

Fig. 21. — Bingham model curve in cartesian coordinates.

If the scale readings are plotted directly as a function of the rotor velocity N in rpm, the plastic viscosity in centipoises will be

$$\mu_p = \frac{0.51(\theta_2 - \theta_1)}{1.7(N_2 - N_1)} \, 1\,000.$$

τ_0 is expressed in lb/100 ft^2.
If $N_2 = 600$ rpm and $N_1 = 300$ rpm : $\mu_p = \theta_2 - \theta_1$ (*cf.* paragr. 1.6.2.1).

3.2.1.2. Graphical determination of parameters for the power-law model

The Fann viscometer readings are plotted in double **logarithmic coordinates,** as in figure 22.

n is proportional to the slope of the best straight line drawn through the four points. K is proportional to the ordinate value of the point $\log \dot{\gamma} = 0$ ($\dot{\gamma} = 1$) on the abscissa.

If the logarithms of scale readings are plotted against the logarithms of the rotor angular velocity expressed in rpm, n will be given by

$$n = \frac{\log \theta_2 - \log \theta_1}{\log N_2 - \log N_1} = \frac{\log \theta_2/\theta_1}{\log N_2/N_1}.$$

K, expressed in lb . s^n/100 ft^2, is the ordinate of the point corresponding to $N = 0.6$ ($\log N = \log 0.6$).

If $N_2 = 600$ rpm and $N_1 = 300$ rpm, we have

$$n = \frac{\log \theta_2/\theta_1}{\log 2} = 3.32 \log \theta_2/\theta_1$$

(*cf*. paragr. 1.6.2.2).

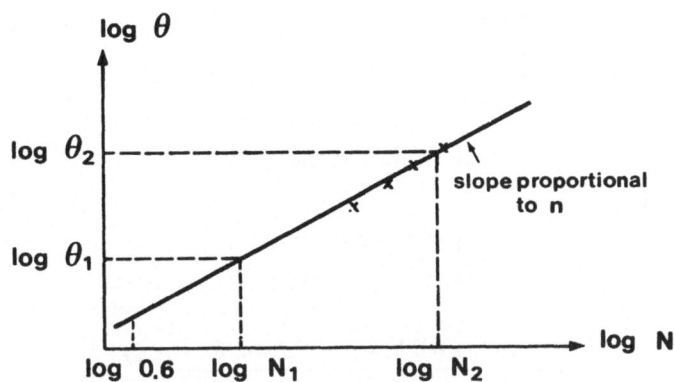

Fig. 22. — Power-law model curve in logarithmic coordinates.

3.2.1.3. The method of least squares

Let P be points whose abscissas are x_p and ordinates are y_p. The straight line fitting the P points may be represented by the equation $y = ax + b$, with the parameters given by

$$a = \frac{\Sigma y_p \Sigma x_p - P\Sigma x_p y_p}{(\Sigma x_p)^2 - P\Sigma x_p^2}$$

$$b = \frac{\Sigma x_p y_p \Sigma x_p - \Sigma y_p \Sigma x_p^2}{(\Sigma x_p)^2 - P\Sigma x_p^2}.$$

If the **Bingham model** has been chosen, the plastic viscosity μ_p and the yield-point τ_0 are deduced from the above equations for a and b by putting $y = \tau$ and $x = \dot{\gamma}$.

In practice, the following equations are used:

(a) In SI units

$$y = 0.51\theta$$
$$x = 1.7N$$

(b) In practical units

$$y = \theta$$
$$x = 1.7N$$

We have

$$\mu_p = 3 \frac{\Sigma\theta\Sigma N - P\Sigma(\theta N)}{(\Sigma N)^2 - P\Sigma N^2} \qquad \text{(in P)}$$

$$\mu_p = 300 \frac{\Sigma\theta\Sigma N - P\Sigma(\theta N)}{(\Sigma N)^2 - P\Sigma N^2} \qquad \text{(in cP)}$$

$$\tau_0 = \frac{\Sigma(N\theta)\Sigma N - \Sigma\theta\Sigma N^2}{(\Sigma N)^2 - P\Sigma N^2} \qquad \text{(in lb/100 ft}^2\text{)}$$

$$\tau_0 = 0.51 \frac{\Sigma(N\theta)\Sigma N - \Sigma\theta\Sigma N^2}{(\Sigma N)^2 - P\Sigma N^2} \qquad \text{(in Pa)}$$

If the **power-law model** has been selected, n and $\log K$ are obtained from the equations for a and b respectively, taking:

(a) In SI units

$$y = \log \tau = \log 0.51\theta$$
$$x = \log \dot{\gamma} = \log 1.7N$$

(b) In practical units

$$y = \log \tau = \log \theta$$
$$x = \log \dot{\gamma} = \log 1.7N$$

We have

$$n = \frac{\Sigma \log \theta \Sigma \log \dot{\gamma} - P\Sigma(\log \theta \cdot \log \dot{\gamma})}{(\Sigma \log \dot{\gamma})^2 - P\Sigma(\log \dot{\gamma})^2}$$

$$\log K = \frac{\Sigma(\log \theta \cdot \log \dot{\gamma})\Sigma \log \dot{\gamma} - \Sigma \log \theta\Sigma(\log \dot{\gamma})^2}{(\Sigma \log \dot{\gamma})^2 - P\Sigma(\log \dot{\gamma})^2} \qquad (K \text{ in lb} \cdot \text{s}^n/100 \text{ ft}^2)$$

The calculations may be considerably facilitated by using a mini-computer.

NOTE.   The six-speed Fann viscometer has certain draw-backs. There is a gap between the 6 rpm and the 100 rpm speeds, i.e., in the shear-rate range between 10 and 170 s^{-1} — which is of the same order as the rates actually measured in the annulus.

3.2.2.   Two-speed Fann viscometer

The choice of the model is difficult; one might be guided by the composition of the fluid. Thus:

(a) Muds with a high content of solids and heavy muds (specific gravity higher than 1.5) behave roughly as Bingham fluids.

(b) Muds with a low solid content (as defined by Ricard in "*Forages*" No 69, pp. 67-95, 1975) behave roughly as power-law fluids.

(c) The addition of a thinning agent modifies the rheological behavior, which becomes closer to that of a power-law fluid.

Clearly, these are mere rules-of-thumb, but nevertheless, their use renders the selection process somewhat less arbitrary.

If the **Bingham model** is chosen, the plastic viscosity μ_p and the yield-point τ_0 are determined from the Fann readings at 300 and 600 rpm, according to Eqs. 16, 17, 18 and 18 *bis* in Chapter 1:

$$\mu_p = \theta_{600} - \theta_{300} \qquad \text{(cP)}$$
$$\tau_0 = \theta_{600} - 2(\theta_{600} - \theta_{300}) \qquad \text{(lb/100 ft}^2\text{)}$$

If the **power-law model** is selected, the index of rheological behavior n and the consistency index K are obtained from Fann readings at 300 and 600 rpm, using Eqs. 23 and 24 in Chapter 1:

$$n = 3.32 \log \frac{\theta_{600}}{\theta_{300}}$$

$$K \text{ (lb . s}^n/100 \text{ ft}^2) = \frac{\theta_{600}}{1\,020^n}$$

3.3. DETERMINATION OF FLOW TYPE AND CALCULATION OF PRESSURE LOSSES

For the relevant equations see the following tables at the end of this Chapter:

Table XIII.   Newtonian fluids. Circulation in drill pipe.

Table XIV.   Newtonian fluids. Circulation in annulus.

Table XV.   Bingham fluids. Circulation in drill pipe.

Table XVI.   Bingham fluids. Circulation in annulus.

Table XVII.   Bingham fluids. Circulation in surface installations.

Table XVIII. Power-law fluids. Circulation in drill pipe.

Table XIX.   Power-law fluids. Circulation in annulus.

Table XX.   Power-law fluids. Circulation in surface installations.

Table XXI.   Pressure loss through bit nozzles.

It will be recalled that pressure losses are calculated by different equations, depending on whether the flow is laminar or turbulent. Thus, as can be seen from the flowchart below, all calculations of pressure losses must be preceded by a determination of the type of flow.

These equations must be used for each section of constant diameter (internal diameter D for drill string; external diameter of drill string D_i and internal diameter of well D_o for the annulus).

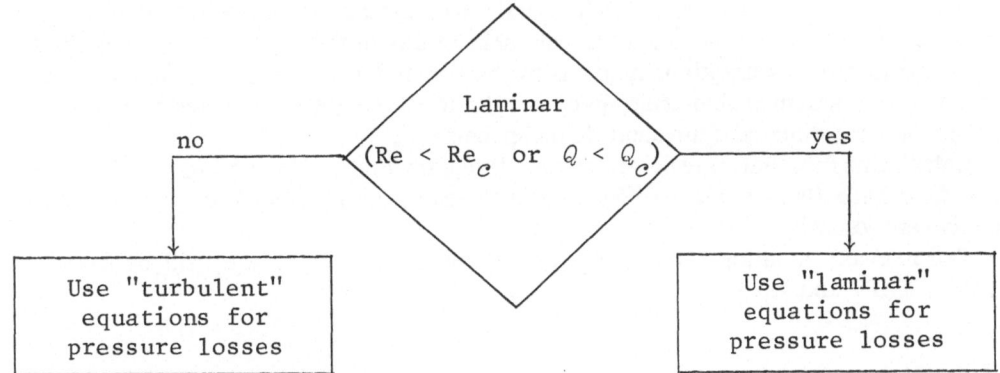

3.4. PRESSURE VARIATIONS DURING EQUIPMENT RUNNING OR PULLING

Table XXII should be used to calculate:

(a) Surge pressure corresponding to gel-breakup.
(b) Pressure variations due to inertial effects.

Calculations of pressure variations due to the displacement of the mud during the operations are carried out according to the flowchart No 1.
For running drill string, the sign of ΔP_a is positive (surge pressure).
For pulling drill string, the sign of ΔP_a is negative (swab pressure).

3.5. EQUIVALENT DENSITY AND EQUIVALENT SPECIFIC GRAVITY IN CIRCULATION

These values are calculated using Table XXIII. If practical units used by French organizations are employed, density and specific gravity have the same numerical values.

3.6. LIFT OF CUTTINGS

It was seen in Chapter 2 that the methods employed for calculating the settling velocity of cuttings in an annulus do not yield strictly accurate results. However, the use of the method which will now be described will yield results of the right order of magnitude, provided that the assumptions made as to the size and shape of the particles are fairly realistic. It is assumed that their shape is cylindrical (height h, diameter d) or roughly spherical, the estimated equivalent diameter being φ_p.

Calculation flowcharts are presented for a Bingham fluid and for a power-law fluid. The two flowcharts (Nos 2 and 3) differ only in the calculation of the flow type and of the equivalent viscosity.

The relevant equations, indicating the system of units employed, will be found in Tables XXIV and XXV.

FLOWCHART 1

CALCULATION OF PRESSURE VARIATIONS DUE TO DISPLACEMENT
OF MUD DURING EQUIPMENT RUNNING OR PULLING

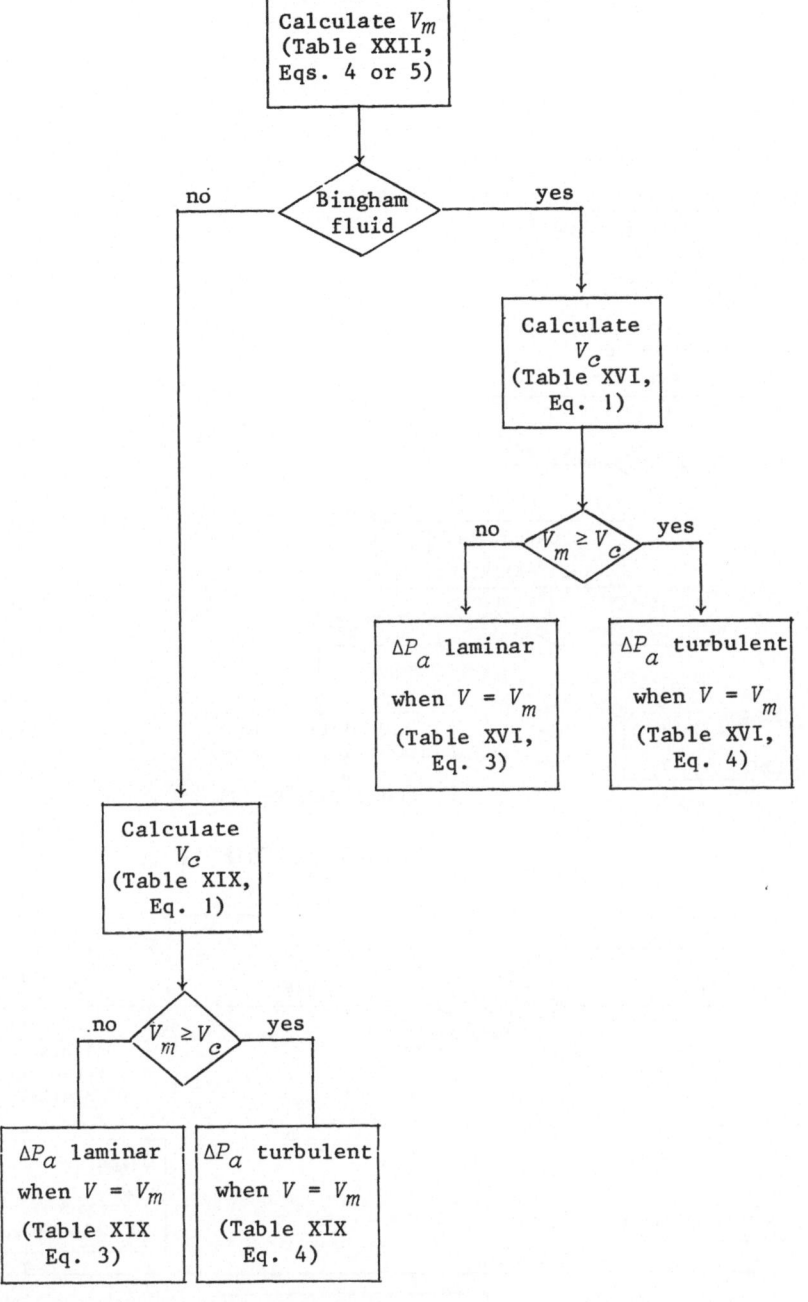

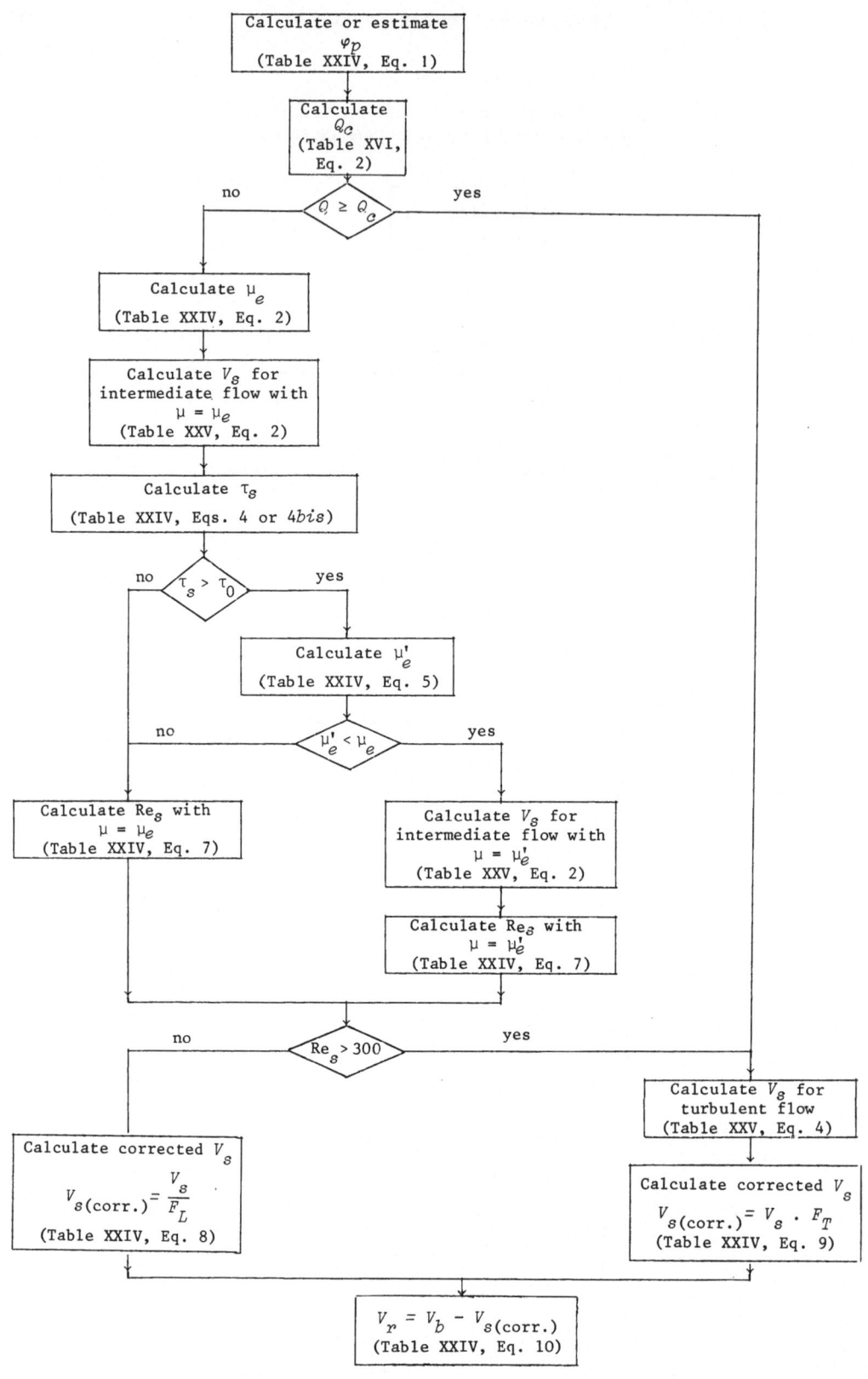

FLOWCHART 2

CALCULATION OF THE LIFT VELOCITY OF CUTTINGS IN A BINGHAM FLUID

FLOWCHART 3

CALCULATION OF THE LIFT VELOCITY OF CUTTINGS
IN A POWER-LAW FLUID

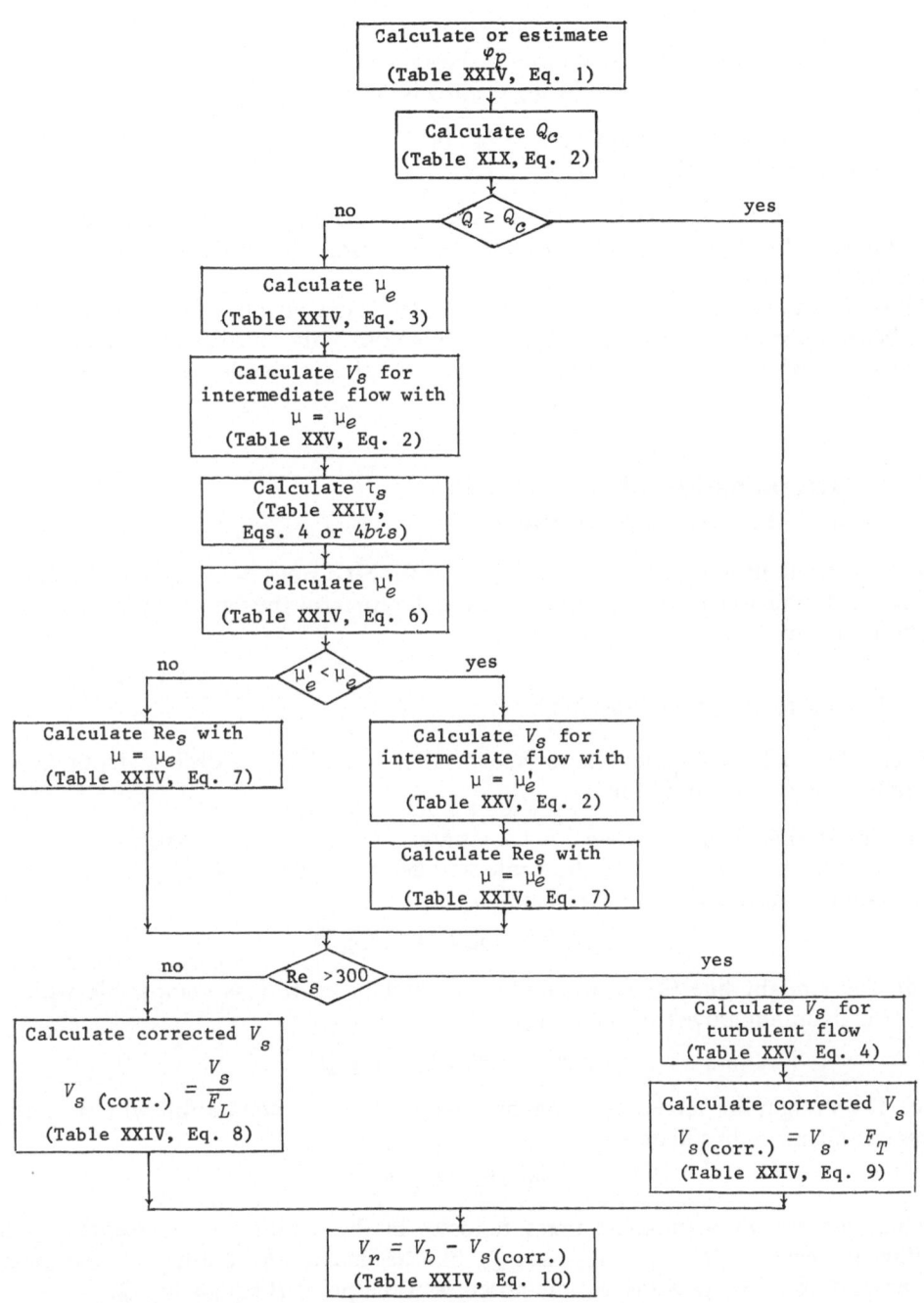

3.7. HYDRAULIC POWER

In principle, the optimum distribution of the available hydraulic power between the bit and the other parts of the circuit during drilling is determined by one of the two criteria:

(a) Maximum hydraulic power at the bit, or
(b) Maximum hydraulic impact.

However, the application of these criteria is not always possible due to external constraints on the flow rate and on the pressure: cleaning of hole, danger of wall erosion, pump liner size, etc.

Thus, the optimum values could not be adopted if the calculated optimum flow rate were below the minimum flow rate Q_m at which the rise of the cuttings could take place at all. Also, at shallow depths, optimum values cannot always be selected because of pump liner size.

3.7.1. Determination of optimum flow rate and pressure loss at the bit

This determination is carried out graphically, for each stage, with the aid of diagrams plotted in double logarithmic coordinates, which represent the various linear equations given in paragr. 2.8.3.

3.7.1.1. Basic diagram (Fig. 23)

Using the available data for the surface installation and for the drilling depth desired, the following graphs are plotted:

(a) The straight lines a_0, a_1, a_2, a_3 ..., representing the pressure losses in the circuit outside the bit as a function of the flow rate, at different depths L_0, L_1, L_2, L_3 ..., where L_0 is the depth corresponding to the beginning of the interval:

$$\log \Delta P_c = \log k_c + m \log Q$$

(b) The straight line (b) representing the total pressure loss compatible with the available hydraulic power of the pump:

$$\log \Delta P_c = \log \mathscr{P}_{h_l} - \log Q$$

(c) The straight line (c) representing the total pressure loss compatible with the service pressure of surface installations:

$$\log \Delta P_l = \log P_s$$

Thus, the overall permissible pressure losses in the circuit are represented by the rectilinear sections AB and BC. Let Q_s be the maximum output of the pump, corresponding to the pressure P_s; the abscissa of the point B equals $\log Q_s$.

The basic diagram also includes the straight line (d) representing the lowest permissible flow rate Q_m:

$$\log Q = \log Q_m$$

The coordinates of point I, which is the point of intersection of lines (c) and (d), are $\log Q_m$ and $\log P_s$.

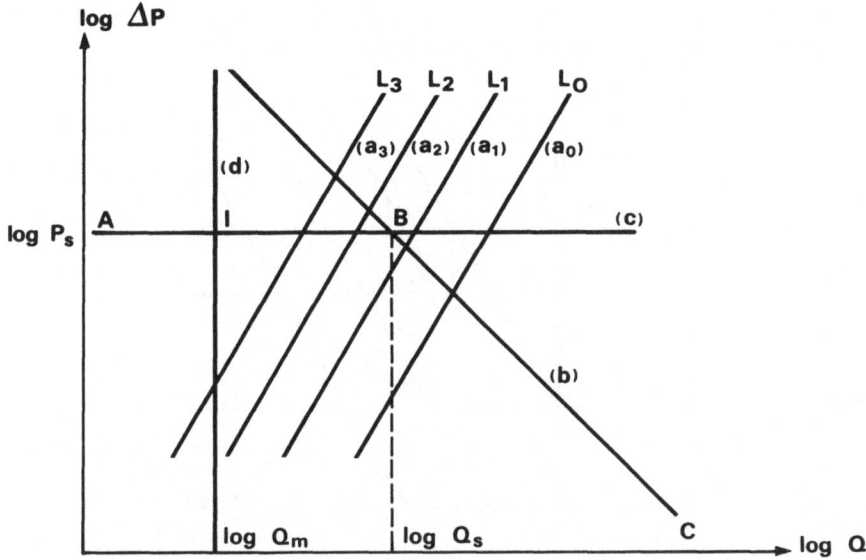

Fig. 23. — Basic diagram.

3.7.1.2. Criterion of maximum hydraulic power at the bit

The procedure is indicated in Figure 24.

If $m = 1.8$, $\dfrac{1}{1 + m} = 0.35$.

In accordance with the results obtained in paragr. 2.8.3.2.A, for Bingham fluid and turbulent flow, the basic diagram is supplemented by marking out the rectilinear section $A_1 B_1$ of line (c'), which is parallel to (c), and is given by the equation

$$\log \Delta P = \log (0.35 P_s)$$

Here, B_1 is the intersection of (c') with the line $\log Q = \log Q_s$, and I_1 is the intersection of (c') with (d).

Let L_1 be the depth corresponding to the line (a_1) through B_1, and let L_3 be the depth corresponding to line (a_3) through I_1.

From the beginning of the interval to depth L_1 (i.e. in zone 1, in which the pressure is not limited by P_s), the optimum flow rate Q_{opt} will be equal to the minimum flow rate Q_m if $Q_m \geqslant Q_s$. The entire hydraulic power $\mathscr{P}_{h_l}$ is utilized.

From depth L_1 to depth L_3 (i.e., in zone 2, in which the pressure is limited by P_s), the following distribution of pressure losses should be considered:

$$\Delta P_c = 0.35\Delta P$$
$$\Delta P_e = 0.65\Delta P$$
$$\Delta P = \Delta P_l$$

The operative hydraulic power at the bit decreases with increasing depth.

The value of the optimum flow rate Q_{opt} is given by the intersections of (c') with lines (a) at each individual depth. As L increases, Q decreases from Q_s at $L = L_1$ to Q_m at $L = L_3$.

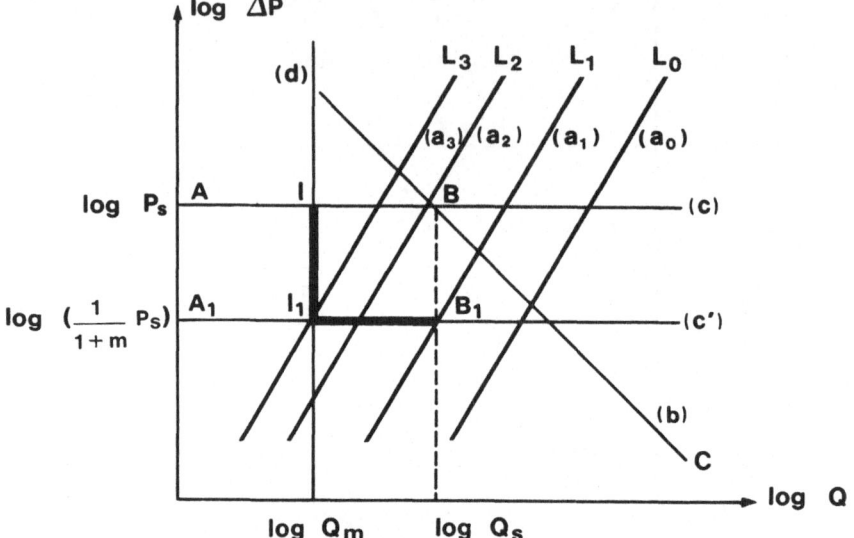

Fig. 24. — Diagrammatic representation of the criterion of maximum hydraulic power at the bit.

Beyond the depth L_3 optimization is no longer possible, and the flow rate stays at the constant value Q_m.

3.7.1.3.　Criterion of maximum hydraulic impact

The method of calculation is as before, using Figure 25.

If $m = 1.8$, $\dfrac{1}{2+m} = 0.26$

$$\dfrac{2}{2+m} = 0.52$$

In accordance with the results obtained in paragr. 2.8.3.2.B, for Bingham fluid and turbulent flow, the lines marked on the basic diagram are:

(a) The rectilinear segments A_2B_2 and B_2C_2, given by

$$\log (0.52\Delta P_l) = f(\log Q)$$

i.e. they are marked on the lines:

. (c''), parallel to (c), given by the equation

$$\log \Delta P = \log (0.52 P_s)$$

. (b'') parallel to (b), given by the equation

$$\log \Delta P = \log (0.52 \mathscr{P}_{h_l}) - \log Q$$

(b) The rectilinear segments $A_3 B_3$ and $B_3 C_3$, given by

$$\log (0.26 \Delta P_l) = f(\log Q)$$

which are marked on the lines:

. (c'''), parallel to (c), given by the equation

$$\log \Delta P = \log (0.26 P_s)$$

. (b''') parallel to (b), given by the equation

$$\log P = \log (0.26 \mathscr{P}_{h_l}) - \log Q$$

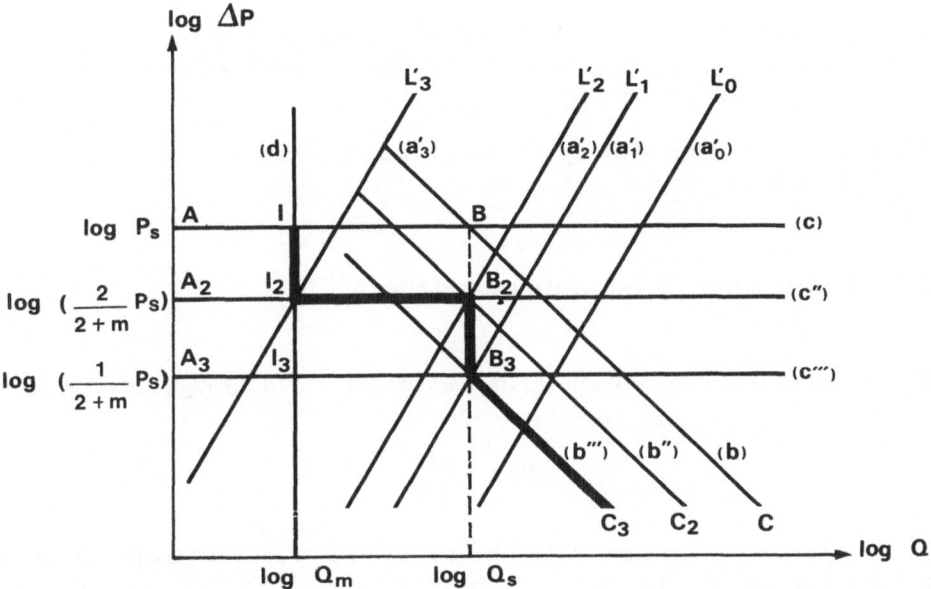

Fig. 25. — Diagrammatic representation of the criterion of maximum hydraulic impact.

B_2 and B_3 are, respectively, the intersections of (c'') with (b'') and of (c''') with (b'''). I_2 is the intersection point of (c'') with (d).

Let:

L'_1 be the depth represented by line (a'_1) through B_3.
L'_2 be the depth represented by line (a'_2) through B_2.
L'_3 be the depth represented by line (a'_3) through I_2.

From the beginning of the interval to depth L'_1 (i.e., in zone 1), the optimum flow rate for each individual depth is given, in theory, by the intersections of line (b''') with lines (a), and should decrease with increasing L.

However, in practical work, adherence to this criterion is not feasible, because one cannot keep changing the pump liner. Nevertheless, an attempt should be made to select a liner which would give a flow rate $Q > Q_s$, depending on the magnitude of L'_1 and on the nature of the drilled strata.

All the available hydraulic power $\mathcal{P}_{h_l}$ is utilized.

Between the depths L'_1 and L'_2 the flow rate is kept constant at Q_s. All the available hydraulic power is utilized.

Between the depths L'_2 and L'_3 (i.e., in zone 2 where pressure limitations obtain), the optimum distribution of pressure losses to be accepted is

$$\Delta P_c = 0.52\Delta P$$
$$\Delta P_e = 0.48\Delta P$$
$$\Delta P = \Delta P_l = P_s$$

and the hydraulic power at the bit decreases with increasing depth.

The optimum rate, Q_{opt}, given by the intersection of line (c'') with lines (a), decreases with increasing L, and becomes equal to Q_m at $L = L'_3$.

Beyond L'_3 optimization of the hydraulic-power distribution is no longer possible, and the flow rate is kept constant at Q_m.

3.7.2. Calculation of bit orifices (nozzles)

The overall surface area available for the passage of the mud through the bit is given by the relation

$$\Delta P_e = k_e \left(\frac{Q}{A}\right)^2$$

(*cf*. Table XXI); the pressure loss at the bit, the working flow rate and the power through-put through the bit are all known.

If this function of Q is plotted in double logarithmic coordinates, a set of straight lines with a slope of 2 is obtained, each line corresponding to a different surface area of the nozzles, as illustrated in Figure 26.

Using the given values of Q and ΔP_e, the size of the nozzles to be used is found from a nomogram. After the optimum combination of nozzles has been determined, it must be tested in order to insure that the rate of flow through them is faster than the minimum permissible value, viz. 6-7.6 m/s (20-25 ft/s) per inch of bit diameter.

The rates thus calculated — usually 120-185 m/s — are, on the average, higher than these limits. If they happen to be lower, it is preferable to use an ordinary rather than a jet bit.

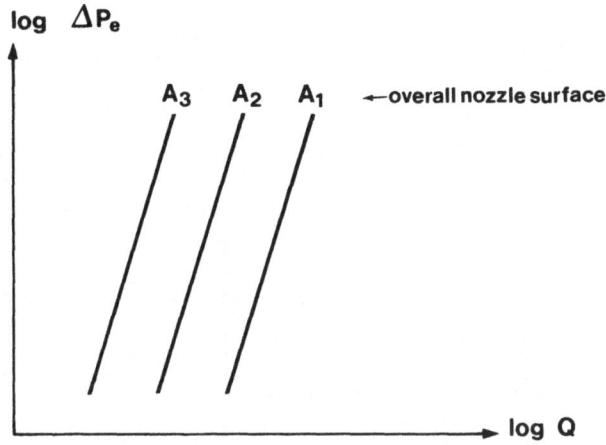

Fig. 26. — Variation of ΔP_e with Q.

3.7.3.   Rate of rise in the annulus

Experience has shown that speeds of 20-25 m/min in the annulus are quite realistic. It should be stressed, however, that the rate of removal of the cuttings depends mainly on their velocity in the annulus. If the flow rate decreases, the mud becomes denser because of the increased cuttings content with consequences of:

(a) Danger of overpressure.
(b) Increase in differential pressure and slow-down in the drilling rate.

To conclude, the following velocities in the annulus may be recommended:

(a) in soft formation, $V = 30\text{-}40$ m/min.
(b) in hard formation, $V = 25\text{-}30$ m/min.

3.7.4.   Constraints on the application of the optimum flow rate

It is sometimes necessary to work at a flow rate below the calculated optimum because of:

(a) Danger of erosion of hole wall by drilling mud moving at a high velocity through the annulus. Such an erosion is undesirable if a gage hole is wanted.
(b) High pressure losses in the annulus, which lead to a rise in the equivalent density at the bit depth.

The flow rate ultimately chosen must represent a compromise between a tolerable pressure loss in the annulus and the highest attainable value of the product $\rho Q V$ or $\rho Q V^2$ at the bit depth.

TABLE X

SUMMARY OF VARIATIONS OF DIFFERENT HYDRAULIC PARAMETERS WITH DEPTH.
POSSIBLE VARIATION DURING A DRILLING INTERVAL (BINGHAM FLUID AND TURBULENT FLOW)

CRITERION \ DEPTH	L_0	L'_1	L_1	L'_2	L_2	L_3	L'_3	L_f
Maximum hydraulic power at bit	$Q = Q_{opt} \geqslant Q_s$ $\quad$ $\mathscr{P}_h$ used $= \mathscr{P}_{h_1}$ $\quad$ $\Delta P = \dfrac{\mathscr{P}_{h_1}}{Q}$ $\quad$ $\Delta P_c \nearrow$ $\quad$ $\Delta P_e \searrow$ $\quad$ $A \nearrow$	(constant)		$Q = Q_{opt} \searrow$ $\;(Q = Q_s$ if $L = L_1$; $\;Q = Q_m$ if $L = L_3)$ $\quad$ $\mathscr{P}_h$ used $\searrow$ $\quad$ $\Delta P = P_s$ $\quad$ $\left.\begin{array}{l}\Delta P_c = 0.35 P_s\\ \Delta P_e = 0.65 P_s\end{array}\right\}$ constant $\quad$ $A \searrow$			$Q = Q_m$ (constant) $\quad$ $\mathscr{P}_h$ used $= P_s Q_m$ $\quad$ $\Delta P = P_s$ $\quad$ $\Delta P_c \nearrow$ $\quad$ $\Delta P_e \searrow$ $\quad$ $A \nearrow$	
Maximum hydraulic impact on the bit	$Q \geqslant Q_s$ $\quad$ $\mathscr{P}_h$ used $= \mathscr{P}_{h_1}$ $\quad$ $\Delta P = \dfrac{\mathscr{P}_{h_1}}{Q}$	$Q = Q_{opt} = Q_s$ (constant) $\quad$ $\mathscr{P}_h$ used $= \mathscr{P}_{h_1}$ $\quad$ $\Delta P = \dfrac{\mathscr{P}_{h_1}}{Q_s} = P_s$ $\quad$ $\Delta P_c \nearrow (\Delta P_c = 0.26 P_s$ if $L = L'_1$; $\Delta P_c = 0.52 P_s$ if $L = L'_2)$ $\quad$ $\Delta P_e \searrow (\Delta P_e = 0.74 P_s$ if $L = L'_1$; $\Delta P_e = 0.48 P_s$ if $L = L_2)$ $\quad$ $A \nearrow$			$Q = Q_{opt} \searrow$ $\;(Q = Q_s$ if $L = L'_2$; $\;Q = Q_m$ if $L = L'_3)$ $\quad$ $\mathscr{P}_h$ used $\searrow$ $\quad$ $\Delta P = P_s$ $\quad$ $\left.\begin{array}{l}\Delta P_c = 0.52 P_s\\ \Delta P_e = 0.48 P_s\end{array}\right\}$ constant $\quad$ $A \searrow$		$Q = Q_m$ (constant) $\quad$ $\mathscr{P}_h$ used $= P_s Q_m$ $\quad$ $\Delta P = P_s$ $\quad$ $\Delta P_c \nearrow$ $\quad$ $\Delta P_e \searrow$ $\quad$ $A \nearrow$	

Zones in which the optimization criteria are valid are framed by heavy lines.

3.7.5   Summary table of variations of different hydraulic parameters with depth

If the distance between L_0 (beginning of interval) and L_f (final depth) is included in all zones of the diagram, Table X should be consulted. Otherwise, the locations of L_0 and L_f with respect to the individual depths L_1, L'_1, L'_2, L_3, L'_3 should be considered.

It is seen that different values may be obtained, depending on whether the power criterion or the impact criterion is used.

3.7.6.   A practical method for the determination of the optimum distribution of hydraulic power at a given depth

The two flowcharts Nos 4 and 5 describe:

(a) A routine based on the criterion of the maximum hydraulic power at the bit.
(b) A routine based on the criterion of maximum hydraulic impact.

The following data are used:

(a) Maximum pump capacity.
(b) Service pressure P_s in the surface installation.
(c) Dimensions of hole and drill string.
(d) Mud parameter (ρ, μ_p, τ_0).
(e) Lower limit Q_m of the flow rate, at which the cuttings will still rise in the annulus.
(f) Upper limit Q_l of the flow rate imposed by the danger of erosion.

It is assumed that:

(a) The mud is a Bingham fluid in turbulent flow.
(b) The calculated flow rates are not limited by pump liner size.
(c) The nozzle velocities obtained from the calculated values of A and Q are sufficiently high.
(d) The limit flow rate Q_l is higher than the minimum rate Q_m. If this is not so, the actual flow rate must be found without using the flowchart, by a compromise between the rising rate of the cuttings and the danger of erosion of the hole wall.

These calculations are made by using the equations given in Table XXVI.

FLOWCHART 4
METHOD BASED ON THE CRITERION OF MAXIMUM
HYDRAULIC POWER AT THE BIT

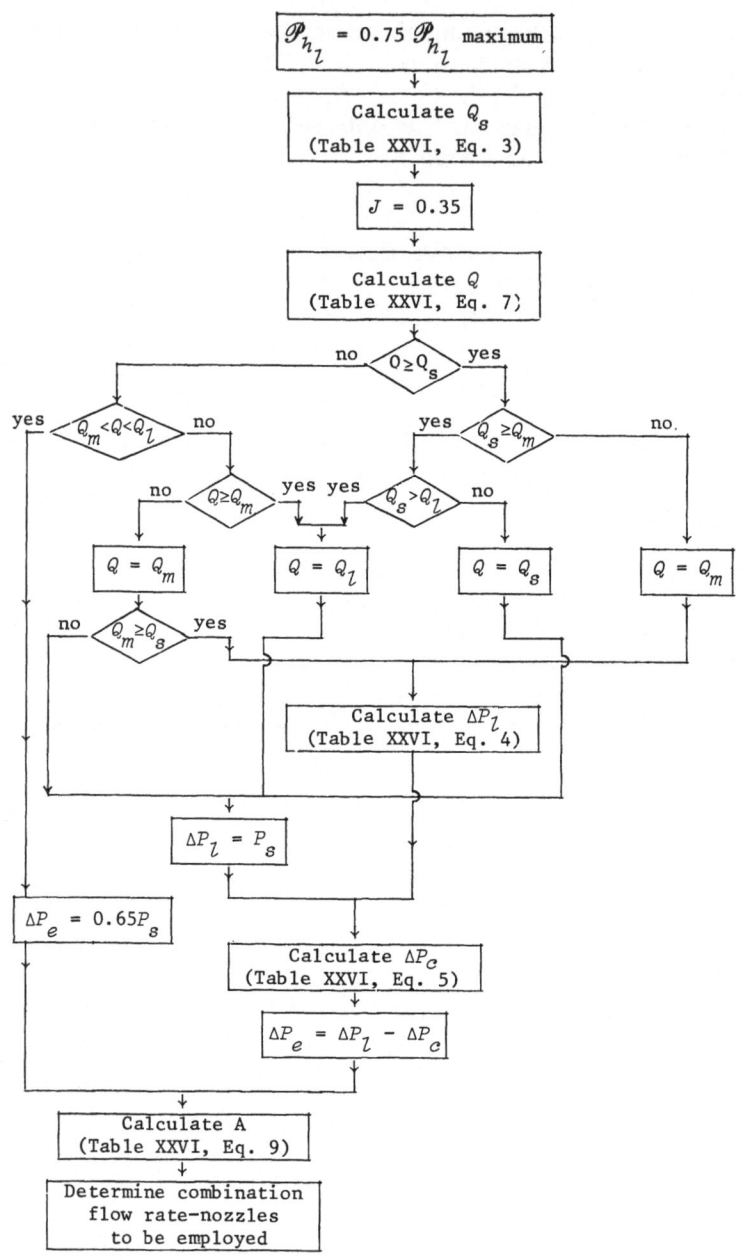

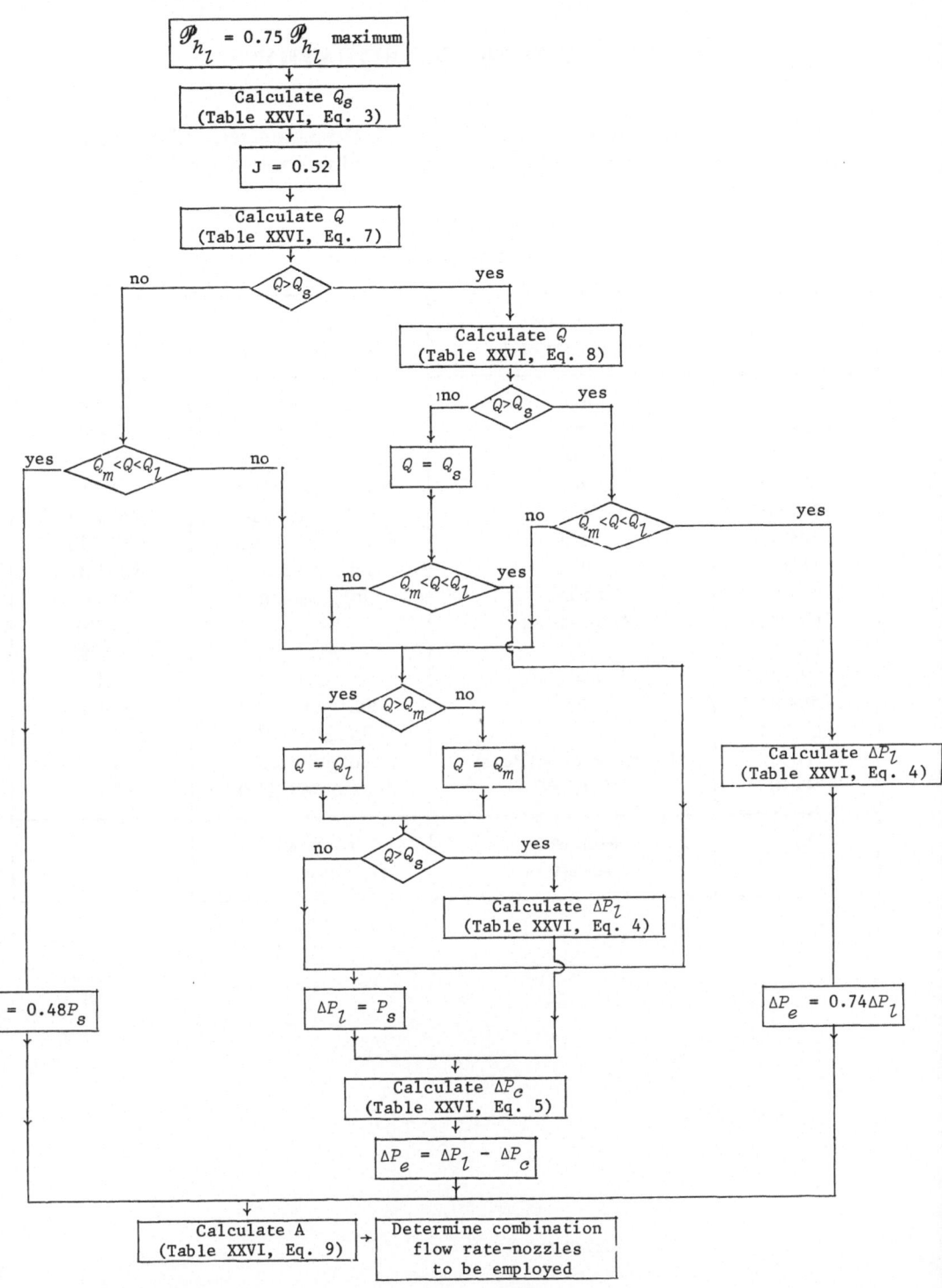

FLOWCHART 5
METHOD BASED ON THE CRITERION OF THE MAXIMUM HYDRAULIC IMPACT

$\mathcal{P}_{h_l} = 0.75 \, \mathcal{P}_{h_l} \text{ maximum}$

Calculate Q_s
(Table XXVI, Eq. 3)

$J = 0.52$

Calculate Q
(Table XXVI, Eq. 7)

$Q > Q_s$ — no / yes

Calculate Q
(Table XXVI, Eq. 8)

$Q > Q_s$ — no / yes

$Q = Q_s$

$Q_m < Q < Q_l$ — yes / no

$Q_m < Q < Q_l$ — no / yes

$Q_m < Q < Q_l$ — no / yes

$Q > Q_m$ — yes / no

$Q = Q_l$ $Q = Q_m$

Calculate ΔP_l
(Table XXVI, Eq. 4)

$Q > Q_s$ — no / yes

Calculate ΔP_l
(Table XXVI, Eq. 4)

$P_e = 0.48 P_s$

$\Delta P_l = P_s$

$\Delta P_e = 0.74 \Delta P_l$

Calculate ΔP_c
(Table XXVI, Eq. 5)

$\Delta P_e = \Delta P_l - \Delta P_c$

Calculate A
(Table XXVI, Eq. 9)

Determine combination
flow rate-nozzles
to be employed

3.8. TABLES OF EQUATIONS

The unit conversion factors are listed in Table XI. The parameters and the equations to be used in the calculations described above are listed in Table XII à XXVI.

TABLE XI

UNIT CONVERSION FACTORS

	TO OBTAIN THE NUMBER OF ↓	MULTIPLY THE NUMBER OF ↓	BY ↓
0.03937	mm	inch (in)	25.4
3.28084	m	foot (ft)	0.3048
0.155	cm^2	square inch (in^2)	6.45163
196.8504	m/s	foot/minute (ft/min)	$5.08 \ 10^{-3}$
11811.02	m/s	foot/hour (ft/h)	$8.46667 \ 10^{-5}$
60000	m^3/s	litre/minute (l/min)	$1.6667 \ 10^{-5}$
15850.37	m^3/s	gallon/minute	$6.309 \ 10^{-5}$
$8.3452 \ 10^{-3}$	kg/m^3	lb_{mass}/gallon	119.829
0.1020	Newton (N)	kg_{force}	9.807
0.22481	N	lb_{force}	4.4482
10^{-5}	Pascal (Pa)	bar	10^5
$1.4504 \ 10^{-4}$	Pa	lb_{force}/in^2 (psi)	6894.7
2.08784	Pa	$lb_{force}/100 \ ft^2$	0.478964
1000	Pascal.seconde (Pa.s)	centipoise (cP)	10^{-3}
$1.341 \ 10^{-3}$	Watt (W)	horsepower (HP)	745.7
BY ↑	MULTIPLY THE NUMBER OF ↑	TO OBTAIN THE NUMBER OF ↑	

TABLE XII

SYMBOLS AND UNITS EMPLOYED IN CHAPTERS 3 AND 4

SYMBOL	MEANING	DIMENSIONS	SI UNITS	AMERICAN UNITS	OTHER UNITS
A	Total nozzle surface area	L^2	m^2	in^2	in^2
a_p	Acceleration of drill string	LT^{-2}	m/s^2	ft/s^2	m/s^2
Av	Drilling rate	LT^{-1}	m/s	ft/h	m/h
C_e	Nozzle coefficient	none			
C_r	Drag coefficient (settling)	none			
d	Diameter of cylindrical cuttings	L	m	in	mm
D	Inner diameter of drill string	L	m	in	in
D_f	Hole diameter	L	m	in	in
D_i	Inner diameter of annulus (outer diameter of drill string)	L	m	in	in
D_o	Outer diameter of annulus	L	m	in	in
g	Acceleration due to gravity	LT^{-2}	m/s^2 (9.81)	ft/s^2 (32.2)	cm/s^2 (981)
g_t	Gel strength at time t	$ML^{-1}T^{-2}$	Pa	$lb_{force}/100\ ft^2$	$lb_{force}/100\ ft^2$
h	Thickness of cylindrical cuttings	L	m	in	mm
I_h	Hydraulic impact	MLT^{-2}	N	lb_{force}	kg_{force}
K	Consistency index	$ML^{-1}T^{n-2}$	$Pa \cdot s^n$	$lb_{force} \cdot s^n/100\ ft^2$	$lb_{force} \cdot s^n/100\ ft^2$
L	Length	L	m	ft	m
n	Power-law index of flow behavior	none			
P_h	Hydrostatic pressure	$ML^{-1}T^{-2}$	Pa	lb_{force}/in^2	bar
P_i	Discharge pressure	$ML^{-1}T^{-2}$	Pa	lb_{force}/in^2	bar
P_r	Surge or swab pressure due to pipe movement (running or pulling)	$ML^{-1}T^{-2}$	Pa	lb_{force}/in^2	bar

TABLE XII (cont.)

SYMBOLS ANS UNITS EMPLOYED IN CHAPTERS 3 AND 4

SYMBOL	MEANING	DIMENSIONS	SI UNITS	AMERICAN UNITS	OTHER UNITS
P_s	Overall service pressure in mud circuit	$ML^{-1}T^{-2}$	Pa	lb_{force}/in^2	bar
$\mathscr{P}_h$	Hydraulic power	$ML^{-2}T^{-3}$	W	horsepower	cheval-vapeur
$\mathscr{P}_{h_1}$	Maximum available hydraulic power	$ML^{-2}T^{-3}$	W	horsepower	cheval-vapeur
Q	Flow rate	L^3T^{-1}	m³/s	gallon/min	l/min
Q_c	Critical flow rate	L^3T^{-1}	m³/s	gallon/min	l/min
Q_l	Maximum flow rate at which erosion is prevented	L^3T^{-1}	m³/s	gallon/min	l/min
Q_m	Minimum flow rate at which cuttings can be lifted	L^3T^{-1}	m³/s	gallon/min	l/min
Re	Reynolds number	none			
Re$_s$	Reynolds number of cuttings	none			
V	Mud circulation velocity	LT^{-1}	m/s	ft/min	m/min
V_c	Critical velocity	LT^{-1}	m/s	ft/min	m/min
V_m	Equivalent velocity of displacement of mud during pipe movement	LT^{-1}	m/s	ft/min	m/min
V_{mf}	Maximum velocity of displacement of mud using closed drill string	LT^{-1}	m/s	ft/min	m/min
V_{mo}	Maximum velocity of displacement of mud using open drill string and complete filling	LT^{-1}	m/s	ft/min	m/min
V_p	Drill string running speed	LT^{-1}	m/s	ft/min	m/min
V_r	Uplift velocity of cuttings	LT^{-1}	m/s	ft/min	m/min

TABLE XII (end)

SYMBOLS AND UNITS EMPLOYED IN CHAPTERS 3 AND 4

SYMBOL	MEANING	DIMENSIONS	SI UNITS	AMERICAN UNITS	OTHER UNITS
V_s	Slip velocity of cuttings	LT^{-1}	m/s	ft/min	m/min
$\dot{\gamma}$	Shear rate	T^{-1}	1/s	1/s	1/s
ΔP_a	Pressure loss through the annulus	$ML^{-1}T^{-2}$	Pa	lb_{force}/in^2	bar
ΔP_e	Pressure loss through the bit	$ML^{-1}T^{-2}$	Pa	lb_{force}/in^2	bar
ΔP_f	Variations of pressure during pipe movement due to inertial effects (closed drill string)	$ML^{-1}T^{-2}$	Pa	lb_{force}/in^2	bar
ΔP_i	Pressure loss inside drill string	$ML^{-1}T^{-2}$	Pa	lb_{force}/in^2	bar
ΔP_o	Variations of pressure during pipe movement due to inertial effects (open drill string)	$ML^{-1}T^{-2}$	Pa	lb_{force}/in^2	bar
ΔP_s	Surface pressure loss	$ML^{-1}T^{-2}$	Pa	lb_{force}/in^2	bar
μ	Dynamic viscosity	$ML^{-1}T^{-1}$	Pa . s	centipoise	centipoise
μ_p	Plastic viscosity	$ML^{-1}T^{-1}$	Pa . s	centipoise	centipoise
μ_e	Equivalent viscosity	$ML^{-1}T^{-1}$	Pa . s	centipoise	centipoise
ρ	Density of mud	ML^{-3}	kg/m^3	$lb_{mass}/gallon$	kg/dm^3
ρ_d	Density of mud containing cuttings	ML^{-3}	kg/m^3	$lb_{mass}/gallon$	kg/dm^3
ρ_s	Density of cuttings	ML^{-3}	kg/m^3	$lb_{mass}/gallon$	kg/dm^3
ρ'	Equivalent circulating density	ML^{-3}	kg/m^3	$lb_{mass}/gallon$	kg/dm^3
τ_0	Yield-point	$ML^{-1}T^{-2}$	Pa	$lb_{force}/100\,ft^2$	$lb_{force}/100\,ft^2$
τ_s	Shear stress resulting from settling of cuttings	$ML^{-1}T^{-2}$	Pa	$lb_{force}/100\,ft^2$	$lb_{force}/100\,ft^2$
φ_P	Equivalent diameter of cuttings	L	m	in	mm

TABLE XIII

Newtonian fluids Circulation in drill pipe

PARAMETERS TO BE CALCULATED	SI UNITS	AMERICAN UNITS	OTHER
(1) V_c	$\dfrac{2\,100\mu}{D\rho}$	$\dfrac{135.82\mu}{D\rho}$	$\dfrac{4.96\mu}{D\rho}$
(2) Q_c	$\dfrac{525\pi D\mu}{\rho}$	$\dfrac{5.54D\mu}{\rho}$	$\dfrac{2.514D\mu}{\rho}$
(3) ΔP_i (laminar)	$\dfrac{32LV\mu}{D^2}$ $\dfrac{128QL\mu}{\pi D^4}$	$\dfrac{VL\mu}{89\,775D^2}$ $\dfrac{QL\mu}{3\,663D^4}$	$\dfrac{VL\mu}{120\,967D^2}$ $\dfrac{QL\mu}{61\,295D^4}$
(4) ΔP_i (turbulent)	$\dfrac{0.1L\rho^{0.8}V^{1.8}\mu^{0.2}}{D^{1.2}}$ $\dfrac{1.2126L\rho^{0.8}Q^{1.8}\mu^{0.2}}{\pi^{1.8}D^{4.8}}$	$\dfrac{L\rho^{0.8}V^{1.8}\mu^{0.2}}{3\,212\,923D^{1.2}}$ $\dfrac{L\rho^{0.8}Q^{1.8}\mu^{0.2}}{10\,141D^{4.8}}$	$\dfrac{L\rho^{0.8}V^{1.8}\mu^{0.2}}{306\,529D^{1.2}}$ $\dfrac{L\rho^{0.8}Q^{1.8}\mu^{0.2}}{90\,163D^{4.8}}$

TABLE XIV

NEWTONIAN FLUIDS. CIRCULATION IN ANNULUS

PARAMETERS TO BE CALCULATED	SI UNITS	AMERICAN UNITS	OTHER UNITS
(1) V_c	$\dfrac{2572\mu}{(D_o - D_i)\rho}$	$\dfrac{166.35\mu}{(D_o - D_i)\rho}$	$\dfrac{6.08\mu}{(D_o - D_i)\rho}$
(2) Q_c	$\dfrac{643\pi(D_o + D_i)\mu}{\rho}$	$\dfrac{6.79(D_o + D_i)\mu}{\rho}$	$\dfrac{3.08(D_o + D_i)\mu}{\rho}$
(3) ΔP_a (laminar)	$\dfrac{48LV\mu}{(D_o - D_i)^2}$ $\dfrac{192QL\mu}{\pi(D_o + D_i)(D_o - D_i)^3}$	$\dfrac{VL\mu}{59\,851(D_o - D_i)^2}$ $\dfrac{QL\mu}{2\,442(D_o + D_i)(D_o - D_i)^3}$	$\dfrac{VL\mu}{80\,645(D_o - D_i)^2}$ $\dfrac{QL\mu}{40\,863(D_o + D_i)(D_o - D_i)^3}$
(4) ΔP_a (turbulent)	$\dfrac{0.1275L\rho^{0.8}V^{1.8}\mu^{0.2}}{(D_o - D_i)^{1.2}}$ $\dfrac{1.5465L\rho^{0.8}Q^{1.8}\mu^{0.2}}{\pi^{1.8}(D_o + D_i)^{1.8}(D_o - D_i)^3}$	$\dfrac{L\rho^{0.8}V^{1.8}\mu^{0.2}}{2\,519\,939(D_o - D_i)^{1.2}}$ $\dfrac{L\rho^{0.8}Q^{1.8}\mu^{0.2}}{7952(D_o + D_i)^{1.8}(D_o - D_i)^3}$	$\dfrac{L\rho^{0.8}V^{1.8}\mu^{0.2}}{240\,415(D_o - D_i)^{1.2}}$ $\dfrac{L\rho^{0.8}Q^{1.8}\mu^{0.2}}{70\,696(D_o + D_i)^{1.8}(D_o - D_i)^3}$

TABLE XV

BINGHAM FLUIDS. CIRCULATION IN DRILL PIPE

PARAMETERS TO BE CALCULATED	SI UNITS	AMERICAN UNITS	OTHER UNITS
(1) V_c	$\dfrac{1050}{D\rho}\left[\mu_p + \sqrt{\mu_p^2 + \dfrac{\tau_0 D^2 \rho}{4\,200}}\right]$	$\dfrac{67.91}{D\rho}\left[\mu_p + \sqrt{\mu_p^2 + 8.816\tau_0 D^2 \rho}\right]$	$\dfrac{2.48}{D\rho}\left[\mu_p + \sqrt{\mu_p^2 + 73.57\tau_0 D^2 \rho}\right]$
(2) Q_c	$\dfrac{262.5\pi D}{\rho}\left[\mu_p + \sqrt{\mu_p^2 + \dfrac{\tau_0 D^2 \rho}{4\,200}}\right]$	$\dfrac{2.77D}{\rho}\left[\mu_p + \sqrt{\mu_p^2 + 8.816\tau_0 D^2 \rho}\right]$	$\dfrac{1.257D}{\rho}\left[\mu_p + \sqrt{\mu_p^2 + 73.57\tau_0 D^2 \rho}\right]$
(3) ΔP_i (laminar)	$\dfrac{32LV\mu_p}{D^2} + \dfrac{4\tau_0 L}{D}$ $\dfrac{128LQ\mu_p}{\pi D^4} + \dfrac{4\tau_0 L}{D}$	$\dfrac{LV\mu_p}{89\,775 D^2} + \dfrac{\tau_0 L}{300D}$ $\dfrac{LQ\mu_p}{3\,663 D^4} + \dfrac{\tau_0 L}{300D}$	$\dfrac{LV\mu_p}{120\,968 D^2} + \dfrac{\tau_0 L}{1\,326D}$ $\dfrac{LQ\mu_p}{61\,295 D^4} + \dfrac{\tau_0 L}{1\,326D}$
(4) ΔP_i (turbulent)	$\dfrac{0.1L\rho^{0.8} V^{1.8}\mu_p^{0.2}}{D^{1.2}}$ $\dfrac{1.2126L\rho^{0.8} Q^{1.8}\mu_p^{0.2}}{\pi^{1.8} D^{4.8}}$	$\dfrac{L\rho^{0.8} V^{1.8}\mu_p^{0.2}}{3\,212\,923 D^{1.2}}$ $\dfrac{L\rho^{0.8} Q^{1.8}\mu_p^{0.2}}{10\,141 D^{4.8}}$	$\dfrac{L\rho^{0.8} V^{1.8}\mu_p^{0.2}}{306\,529 D^{1.2}}$ $\dfrac{L\rho^{0.8} Q^{1.8}\mu_p^{0.2}}{90\,163 D^{4.8}}$

TABLE XVI

BINGHAM FLUIDS. CIRCULATION IN ANNULUS

PARAMETERS TO BE CALCULATED	SI UNITS	AMERICAN UNITS	OTHER UNITS
(1) V_c	$\dfrac{1286}{(D_o - D_i)\rho}\left[\mu_p + \sqrt{\mu_p^2 + \dfrac{\tau_0(D_o - D_i)^2\rho}{7716}}\right]$	$\dfrac{83.17}{(D_o - D_i)\rho}\left[\mu_p + \sqrt{\mu_p^2 + 4.8\tau_0(D_o - D_i)^2\rho}\right]$	$\dfrac{3.04}{(D_o - D_i)\rho}\left[\mu_p + \sqrt{\mu_p^2 + 40.05\tau_0(D_o - D_i)^2\rho}\right]$
(2) Q_c	$\dfrac{321.49\pi(D_o + D_i)}{\rho}\left[\mu_p + \sqrt{\mu_p^2 + \dfrac{\tau_0(D_o - D_i)^2\rho}{7716}}\right]$	$\dfrac{3.39(D_o + D_i)}{\rho}\left[\mu_p + \sqrt{\mu_p^2 + 4.8\tau_0(D_o - D_i)^2\rho}\right]$	$\dfrac{1.539(D_o + D_i)}{\rho}\left[\mu_p + \sqrt{\mu_p^2 + 40.05\tau_0(D_o - D_i)^2\rho}\right]$
(3) ΔP_a (laminar)	$\dfrac{48LV\mu_p}{(D_o - D_i)^2} + \dfrac{4\tau_0 L}{D_o - D_i}$ $\dfrac{192LQ\mu_p}{\pi(D_o + D_i)(D_o - D_i)^3} + \dfrac{4\tau_0 L}{D_o - D_i}$	$\dfrac{LV\mu_p}{59851(D_o - D_i)^2} + \dfrac{\tau_0 L}{300(D_o - D_i)}$ $\dfrac{LQ\mu_p}{2442(D_o + D_i)(D_o - D_i)^3} + \dfrac{\tau_0 L}{300(D_o - D_i)}$	$\dfrac{LV\mu_p}{80645(D_o - D_i)^2} + \dfrac{\tau_0 L}{1326(D_o - D_i)}$ $\dfrac{LQ\mu_p}{40863(D_o + D_i)(D_o - D_i)^3} + \dfrac{\tau_0 L}{1326(D_o - D_i)}$
(4) ΔP_a (turbulent)	$\dfrac{0.1275L\rho^{0.8}V^{1.8}\mu_p^{0.2}}{(D_o - D_i)^{1.2}}$ $\dfrac{1.5465L\rho^{0.8}Q^{1.8}\mu_p^{0.2}}{\pi^{1.8}(D_o + D_i)^{1.8}(D_o - D_i)^3}$	$\dfrac{L\rho^{0.8}V^{1.8}\mu_p^{0.2}}{2519939(D_o - D_i)^{1.2}}$ $\dfrac{L\rho^{0.8}Q^{1.8}\mu_p^{0.2}}{7952(D_o + D_i)^{1.8}(D_o - D_i)^3}$	$\dfrac{L\rho^{0.8}V^{1.8}\mu_p^{0.2}}{240415(D_o - D_i)^{1.2}}$ $\dfrac{L\rho^{0.8}Q^{1.8}\mu_p^{0.2}}{70696(D_o + D_i)^{1.8}(D_o - D_i)^3}$

TABLE XVII

BINGHAM FLUIDS. CIRCULATION IN SURFACE INSTALLATIONS

EQUIPEMENT TYPE	AMERICAN UNITS	OTHER UNITS
I. Standpipe 40 ft . 3 in ID Rotary hose 40 ft . 2 in ID Swivel 4 ft . 2 in ID Kelly 40 ft . $2^1/_4$ in ID	$\Delta P_s = 2.525 \cdot 10^{-4} \rho^{0.8} Q^{1.8} \mu_p^{0.2}$	$\Delta P_s = 8.790 \cdot 10^{-6} \rho^{0.8} Q^{1.8} \mu_p^{0.2}$
II. Standpipe 40 ft . $3^1/_2$ in ID Rotary hose 55 ft . $2^1/_2$ in ID Swivel 5 ft . $2^1/_2$ in ID Kelly 40 ft . $3^1/_4$ in ID	$\Delta P_s = 9.619 \cdot 10^{-5} \rho^{0.8} Q^{1.8} \mu_p^{0.2}$	$\Delta P_s = 3.298 \cdot 10^{-6} \rho^{0.8} Q^{1.8} \mu_p^{0.2}$
III. Standpipe 45 ft . 4 in ID Rotary hose 55 ft . 3 in ID Swivel 5 ft . $2^1/_2$ in ID Kelly 40 ft . $3^1/_4$ in ID	$\Delta P_s = 5.335 \cdot 10^{-5} \rho^{0.8} Q^{1.8} \mu_p^{0.2}$	$\Delta P_s = 1.829 \cdot 10^{-6} \rho^{0.8} Q^{1.8} \mu_p^{0.2}$
IV. Standpipe 45 ft . 4 in ID Rotary hose 55 ft . 3 in ID Swivel 6 ft . 3 in ID Kelly 40 ft . 4 in ID	$\Delta P_s = 4.163 \cdot 10^{-5} \rho^{0.8} Q^{1.8} \mu_p^{0.2}$	$\Delta P_s = 1.427 \cdot 10^{-6} \rho^{0.8} Q^{1.8} \mu_p^{0.2}$

TABLE XVIII
Power-law fluids. Circulation in drill pipe

PARAMETERS TO BE CALCULATED	SI UNITS	AMERICAN UNITS	OTHER UNITS
(1) V_c	$\left[\dfrac{(3470-1370n)K}{8\rho}\right]^{\frac{1}{2-n}}\left[\dfrac{6n+2}{Dn}\right]^{\frac{n}{2-n}}$	$1.969\left[\dfrac{5(3470-1370n)K}{\rho}\right]^{\frac{1}{2-n}}\left[\dfrac{3n+1}{1.27Dn}\right]^{\frac{n}{2-n}}$	$0.6\left[\dfrac{(3470-1370n)K}{1.27\rho}\right]^{\frac{1}{2-n}}\left[\dfrac{3n+1}{1.27Dn}\right]^{\frac{n}{2-n}}$
(2) Q_c	$\left[\dfrac{(3470-1370n)\pi^2 D^4 K}{128\rho}\right]^{\frac{1}{2-n}}\left[\dfrac{24n+8}{\pi D^3 n}\right]^{\frac{n}{2-n}}$	$0.016\left[\dfrac{128(3470-1370n)D^4 K}{\rho}\right]^{\frac{1}{2-n}}\left[\dfrac{3n+1}{6.44D^3 n}\right]^{\frac{n}{2-n}}$	$0.06\left[\dfrac{15.37(3470-1370n)D^4 K}{\rho}\right]^{\frac{1}{2-n}}\left[\dfrac{3n+1}{6.44D^3 n}\right]^{\frac{n}{2-n}}$
(3) ΔP_i (laminar)	$\dfrac{4KL}{D}\left(\dfrac{V}{D}\dfrac{6n+2}{n}\right)^n$ $\dfrac{4KL}{D}\left(\dfrac{Q}{\pi D^3}\dfrac{24n+8}{n}\right)^n$	$\dfrac{KL}{300D}\left(\dfrac{0.4V}{D}\dfrac{3n+1}{n}\right)^n$ $\dfrac{KL}{300D}\left(\dfrac{9.8Q}{D^3}\dfrac{3n+1}{n}\right)^n$	$\dfrac{KL}{1326D}\left(\dfrac{1.31V}{D}\dfrac{3n+1}{n}\right)^n$ $\dfrac{KL}{1326D}\left(\dfrac{2.59Q}{D^3}\dfrac{3n+1}{n}\right)^n$
(4) ΔP_i (turbulent)	$\dfrac{(\log n+2.5)\rho V^2 L}{25D}\left[\dfrac{K\left(\dfrac{V}{D}\dfrac{6n+2}{n}\right)^n}{8\rho V^2}\right]^{\frac{1.4-\log n}{7}}$ $\dfrac{32[\log n+2.5)\rho Q^2 L}{50\pi^2 D^5}\left[\dfrac{\pi^2 D^4 K\left(\dfrac{Q}{\pi D^3}\dfrac{24n+8}{n}\right)^n}{128Q^2\rho}\right]^{\frac{1.4-\log n}{7}}$	$\dfrac{(\log n+2.5)\rho V^2 L}{4\,645\,029}\left[\dfrac{19.36K\left(\dfrac{0.4V}{D}\dfrac{3n+1}{n}\right)^n}{\rho V^2}\right]^{\frac{1.4-\log n}{7}}$ $\dfrac{(\log n+2.5)\rho Q^2 L}{7732D^5}\left[\dfrac{D^4 K\left(\dfrac{9.8Q}{D^3}\dfrac{3n+1}{n}\right)^n}{31.03Q^2\rho}\right]^{\frac{1.4-\log n}{7}}$	$\dfrac{(\log n+2.5)\rho V2L}{228\,600D}\left[\dfrac{K\left(\dfrac{1.31V}{D}\dfrac{3n+1}{n}\right)^n}{4.64\rho V^2}\right]^{\frac{1.4-\log n}{7}}$ $\dfrac{(\log n+2.5)\rho Q^2 L}{58\,694D^5}\left[\dfrac{D^4 K\left(\dfrac{2.59Q}{D^3}\dfrac{3n+1}{n}\right)^n}{18.07Q^2\rho}\right]^{\frac{1.4-\log n}{7}}$

TABLE XIX

POWER-LAW FLUIDS. CIRCULATION IN ANNULUS

(Equations in SI units)

PARAMETERS TO BE CALCULATED	SI UNITS
(1) V_c	$\left[\dfrac{(3470 - 1370n)K}{12 \cdot 0.8165\rho}\right]^{\frac{1}{2-n}} \left[\dfrac{8n+4}{(D_o-D_i)n}\right]^{\frac{n}{2-n}}$
(2) Q_c	$\left[\dfrac{\pi^2(3470 - 1370n)(D_o+D_i)^2(D_o-D_i)^2 K}{12 \cdot 16 \cdot 0.8165\rho}\right]^{\frac{1}{2-n}} \cdot \left[\dfrac{32n+16}{\pi(D_o+D_i)(D_o-D_i)^2 n}\right]^{\frac{n}{2-n}}$
(3) ΔP_a (laminar)	$\dfrac{4KL}{D_o-D_i}\left(\dfrac{V}{D_o-D_i}\;\dfrac{8n+4}{n}\right)^n$ $\dfrac{4KL}{D_o-D_i}\left(\dfrac{Q}{\pi(D_o+D_i)(D_o-D_i)^2}\;\dfrac{32n+16}{n}\right)^n$
(4) ΔP_a (turbulent)	$\dfrac{(\log n + 2.5)\rho V^2 L}{25 \cdot 0.8165(D_o-D_i)}\left[\dfrac{K\left(\dfrac{V}{D_o-D_i}\;\dfrac{8n+4}{n}\right)^n}{12 \cdot 0.8165\rho V^2}\right]^{\frac{1.4-\log n}{7}}$ $\dfrac{32(\log n + 2.5)\rho Q^2 L}{50 \cdot 0.8165\pi^2(D_o+D_i)^2(D_o-D_i)^3}\left[\dfrac{\pi^2(D_o+D_i)^2(D_o-D_i)^2 K\left(\dfrac{Q}{\pi(D_o+D_i)(D_o-D_i)^2}\;\dfrac{32n+16}{n}\right)^n}{192 \cdot 0.8165 Q^2\rho}\right]^{\frac{1.4-\log n}{7}}$

TABLE XIX (cont.)

POWER-LAW FLUIDS. CIRCULATION IN ANNULUS

(Equations in American units)

PARAMETERS TO BE CALCULATED	AMERICAN UNITS
(1) V_c	$1.969\left[\dfrac{4.08(3470-1370n)K}{\rho}\right]^{\frac{1}{2-n}}\left[\dfrac{2n+1}{0.64(D_o-D_i)n}\right]^{\frac{n}{2-n}}$
(2) Q_c	$0.016\left[\dfrac{104.74(3470-1370n)(D_o+D_i)^2(D_o-D_i)^2K}{\rho}\right]^{\frac{1}{2-n}}\left[\dfrac{2n+1}{3.22(D_o+D_i)(D_o-D_i)^2n}\right]^{\frac{n}{2-n}}$
(3) ΔP_a (laminar)	$\dfrac{KL}{300(D_o-D_i)}\left(\dfrac{0.8V}{D_o-D_i}\dfrac{2n+1}{n}\right)^n$ $\dfrac{KL}{300(D_o-D_i)}\left(\dfrac{19.6Q}{(D_o+D_i)(D_o-D_i)^2}\dfrac{2n+1}{n}\right)^n$
(4) ΔP_a (turbulent)	$\dfrac{\left[15.81K\left(\dfrac{0.8V}{D_o-D_i}\dfrac{2n+1}{n}\right)^n\right]^{\frac{1.4-\log n}{7}}(\log n+2.5)\rho V^2L}{3\,792\,669(D_o-D_i)}$ $\dfrac{\left[(D_o+D_i)^2(D_o-D_i)^2K\left(\dfrac{19.6Q}{(D_o+D_i)(D_o-D_i)^2}\dfrac{2n+1}{n}\right)^n\dfrac{1}{38\rho Q^2}\right]^{\frac{1.4-\log n}{7}}(\log n+2.5)\rho Q^2L}{6314(D_o+D_i)^2(D_o-D_i)^3}$

TABLE XIX (end)

POWER-LAW FLUIDS. CIRCULATION IN ANNULUS

(Equations in other units)

PARAMETERS TO BE CALCULATED	OTHER UNITS
(1) V_c	$0.6\left[\dfrac{(3470-1370n)K}{2.05\rho}\right]^{\frac{1}{2-n}}\left[\dfrac{2n+1}{0.64(D_o-D_i)n}\right]^{\frac{n}{2-n}}$
(2) Q_c	$0.06\left[\dfrac{12.55(3470-1370n)K(D_o+D_i)^2(D_o-D_i)}{\rho}\right]^{\frac{1}{2-n}}\left[\dfrac{2n+1}{3.22(D_o+D_i)(D_o-D_i)^2 n}\right]^{\frac{n}{2-n}}$
(3) ΔP_a (laminar)	$\dfrac{KL}{1326(D_o-D_i)}\left(\dfrac{2.62V}{D_o-D_i}\dfrac{2n+1}{n}\right)^n$ $\dfrac{KL}{1326(D_o-D_i)}\left(\dfrac{5.18Q}{(D_o+D_i)(D_o-D_i)^2}\dfrac{2n+1}{n}\right)^n$
(4) ΔP_a (turbulent)	$\dfrac{(\log n+2.5)\rho V^2 L}{186\,652(D_o-D_i)}\left[K\dfrac{\left(\dfrac{2.62V}{D_o-D_i}\dfrac{2n+1}{n}\right)^n}{5.68\rho V^2}\right]^{\frac{1.4-\log n}{7}}$ $\dfrac{(\log n+2.5)\rho Q^2 L}{47\,923(D_o+D_i)^2(D_o-D_i)^3}\left[\dfrac{(D_o+D_i)^2(D_o-D_i)^2K\left(\dfrac{5.18Q}{(D_o+D_i)(D_o-D_i)^2}\dfrac{2n+1}{n}\right)^n}{22.13Q^2\rho}\right]^{\frac{1.4-\log n}{7}}$

TABLE XX

POWER-LAW FLUIDS. CIRCULATION IN SURFACE INSTALLATIONS
(Equations in American units)

EQUIPMENT TYPE	AMERICAN UNITS
I. Standpipe 40 ft . 3 in ID Rotary hose 40 ft . 2 in ID Swivel 4 ft . 2 in ID Kelly 40 ft . $2^{1}/_{4}$ in ID	$\Delta P_s = 2.888 \cdot 10^{-4}(\log n + 2.5)\rho Q^2 \left[1.075 K \left(\dfrac{Q \dfrac{3n+1}{n}}{1.416} \right)^n \dfrac{}{Q^2 \rho} \right]^{\frac{1.4 - \log n}{7}}$
II. Standpipe 40 ft . $3^{1}/_{2}$ in ID Rotary hose 55 ft . $2^{1}/_{2}$ in ID Swivel 5 ft . $2^{1}/_{2}$ in ID Kelly 40 ft . $3^{1}/_{4}$ in ID	$\Delta P_s = 1.036 \cdot 10^{-4}(\log n + 2.5)\rho Q^2 \left[2.61 K \left(\dfrac{Q \dfrac{3n+1}{n}}{2.755} \right)^n \dfrac{}{Q^2 \rho} \right]^{\frac{1.4 - \log n}{7}}$
III. Standpipe 45 ft . 4 in ID Rotary hose 55 ft . 3 in ID Swivel 5 ft . $2^{1}/_{2}$ in ID Kelly 40 ft . $3^{1}/_{4}$ in ID	$\Delta P_s = 5.584 \cdot 10^{-5}(\log n + 2.5)\rho Q^2 \left[4.118 K \left(\dfrac{Q \dfrac{3n+1}{n}}{3.878} \right)^n \dfrac{}{Q^2 \rho} \right]^{\frac{1.4 - \log n}{7}}$
IV. Standpipe 45 ft . 4 in ID Rotary hose 55 ft . 3 in ID Swivel 6 ft . 3 in ID Kelly 40 ft . 4 in ID	$\Delta P_s = 4.3197 \cdot 10^{-5}(\log n + 2.5)\rho Q^2 \left[5.307 K \left(\dfrac{Q \dfrac{3n+1}{n}}{4.691} \right)^n \dfrac{}{Q^2 \rho} \right]^{\frac{1.4 - \log n}{7}}$

TABLE XX (end)

POWER-LAW FLUIDS. CIRCULATION IN SURFACE INSTALLATIONS
(Equations in other units)

EQUIPMENT TYPE	OTHER UNITS
I. Standpipe 40 ft . 3 in ID Rotary hose 40 ft . 2 in ID Swivel 4 ft . 2 in ID Kelly 40 ft . $2^{1}/_{4}$ in ID	$\Delta P_s = 1.1597 \cdot 10^{-5}(\log n + 2.5)\rho Q^2 \left[\dfrac{1.846K\left(\dfrac{Q\,\frac{3n+1}{n}}{5.359}\right)^{n}}{Q^2\rho}\right]^{\frac{1.4-\log n}{7}}$
II. Standpipe 40 ft . $3^{1}/_{2}$ in ID Rotary hose 55 ft . $2^{1}/_{2}$ in ID Swivel 5 ft . $2^{1}/_{2}$ in ID Kelly 40 ft . $3^{1}/_{4}$ in ID	$\Delta P_s = 4.159 \cdot 10^{-6}(\log n + 2.5)\rho Q^2 \left[\dfrac{4.483K\left(\dfrac{Q\,\frac{3n+1}{n}}{10.425}\right)^{n}}{Q^2\rho}\right]^{\frac{1.4-\log n}{7}}$
III. Standpipe 45 ft . 4 in ID Rotary hose 55 ft . 3 in ID Swivel 5 ft . $2^{1}/_{2}$ in ID Kelly 40 ft . $3^{1}/_{4}$ in ID	$\Delta P_s = 2.2424 \cdot 10^{-6}(\log n + 2.5)\rho Q^2 \left[\dfrac{7.071K\left(\dfrac{Q\,\frac{3n+1}{n}}{14.673}\right)^{n}}{Q^2\rho}\right]^{\frac{1.4-\log n}{7}}$
IV. Standpipe 45 ft . 4 in ID Rotary hose 55 ft . 3 in ID Swivel 6 ft . 3 in ID Kelly 40 ft . 4 in ID	$\Delta P_s = 1.7347 \cdot 10^{-6}(\log n + 2.5)\rho Q^2 \left[\dfrac{9.112K\left(\dfrac{Q\,\frac{3n+1}{n}}{17.748}\right)^{n}}{Q^2\rho}\right]^{\frac{1.4-\log n}{7}}$

Table XXI
Pressure losses through bit nozzles

SI units	American units	Other units
$\Delta P_e = \dfrac{\rho V^2}{1.975\,C_e^2}$	$\Delta P_e = \dfrac{\rho V^2}{4\,403\,479\,C_e^2}$	$\Delta P_e = \dfrac{\rho V^2}{711\,000\,C_e^2}$

or:

SI units	American units	Other units
$\Delta P_e = \dfrac{\rho Q^2}{1.975\,C_e^2 A^2}$	$\Delta P_e = \dfrac{\rho Q^2}{11\,884\,C_e^2 A^2}$	$\Delta P_e = \dfrac{\rho Q^2}{295\,941\,C_e^2 A^2}$

C_e = Nozzle coefficient.
C_e = 0.80 for a conventional bit.
C_e = 0.95 for a jet bit.

Table XXII
Surge and swab pressure due to drill pipe movements

Parameters to be calculated	SI units	American units	Other units
(1) P_r	$\dfrac{4g_tL}{D_o - D_i}$	$\dfrac{g_tL}{300(D_o - D_i)}$	$\dfrac{g_tL}{1\,326(D_o - D_i)}$
(2) ΔP_f	$\dfrac{L\rho D_i^2 a_p}{D_o^2 - D_i^2}$	$\dfrac{L\rho D_i^2 a_p}{619(D_o^2 - D_i^2)}$	$\dfrac{L\rho D_i^2 a_p}{100(D_o^2 - D_i^2)}$
(3) ΔP_o	$\dfrac{L\rho(D_i^2 - D^2)a_p}{D_o^2 - D_i^2 + D^2}$	$\dfrac{L\rho(D_i^2 - D^2)a_p}{619(D_o^2 - D_i^2 + D^2)}$	$\dfrac{L\rho(D_i^2 - D^2)a_p}{100(D_o^2 - D_i^2 + D^2)}$
(4) V_{mf}	$1.5V_p\left(\dfrac{D_i^2}{D_o^2 - D_i^2} + 0.45\right)$	$1.5V_p\left(\dfrac{D_i^2}{D_o^2 - D_i^2} + 0.45\right)$	$1.5V_p\left(\dfrac{D_i^2}{D_o^2 - D_i^2} + 0.45\right)$
(5) V_{mo}	$1.5V_p\left(\dfrac{D_i^2 - D^2}{D_o^2 - D_i^2 + D^2} + 0.45\right)$	$1.5V_p\left(\dfrac{D_i^2 - D^2}{D_o^2 - D_i^2 + D^2} + 0.45\right)$	$1.5V_p\left(\dfrac{D_i^2 - D^2}{D_o^2 - D_i^2 + D^2} + 0.45\right)$

TABLE XXIII

EQUIVALENT CIRCULATING DENSITY AND EQUIVALENT CIRCULATING SPECIFIC GRAVITY

PARAMETERS TO BE CALCULATED	SI UNITS	AMERICAN UNITS	OTHER UNITS
ρ_d	$\rho + \dfrac{\pi D_f^2 A v (2.5 - \rho)}{4Q}$	$\rho + \dfrac{6.80 \cdot 10^{-4} D_f^2 A v (20.86 - \rho)}{Q}$	$\rho + \dfrac{8.45 \cdot 10^{-3} D_f^2 A v (2.5 - \rho)}{Q}$
P_h	$10 \rho_d L$	$0.0529 \rho_d L$	$\dfrac{\rho_d L}{10}$
ρ'	$\rho_d + \dfrac{\Delta P_a}{10L}$	$\rho_d + \dfrac{18.87 \Delta P_a}{L}$	$\rho_d + \dfrac{10 \Delta P_a}{L}$
Equivalent circulating specific gravity	$\dfrac{\rho_d}{1\,000} + \dfrac{\Delta P_a}{10\,000L}$	$0.119829 \rho_d + \dfrac{2.26 \Delta P_a}{L}$	$\rho_d + \dfrac{10 \Delta P_a}{L}$

Table XXIV
Lift of Cuttings in the Annulus

Parameters to be calculated	SI units	American units	Other units
(1) φ_p (for a disk)	$1.145\sqrt[3]{hd^2}$	$1.145\sqrt[3]{hd^2}$	$1.145\sqrt[3]{hd^2}$
(2) μ_e (Bingham model)	$\mu_p + \dfrac{\pi\tau_0(D_o - D_i)^2(D_o + D_i)}{48Q}$	$\mu_p + \dfrac{8.14\tau_0(D_o - D_i)^2(D_o + D_i)}{Q}$	$\mu_p + \dfrac{30.82\tau_0(D_o - D_i)^2(D_o + D_i)}{Q}$
(3) μ_e (Power-law model)	$K\left(\dfrac{48Q}{\pi(D_o - D_i)^2(D_o + D_i)}\,\dfrac{2n+1}{3n}\right)^{n-1}$	$479K\left(\dfrac{58.82Q}{(D_o - D_i)^2(D_o + D_i)}\,\dfrac{2n+1}{3n}\right)^{n-1}$	$479K\left(\dfrac{15.54Q}{(D_o - D_i)^2(D_o + D_i)}\,\dfrac{2n+1}{3n}\right)^{n-1}$
(4) τ_s (for a disk)	$68.59\sqrt{h(\rho_s - \rho)}$	$7.9\sqrt{h(\rho_s - \rho)}$	$4.53\sqrt{h(\rho_s - \rho)}$
(4 bis) τ_s (for a sphere)	$56\sqrt{d(\rho_s - \rho)}$	$6.45\sqrt{d(\rho_s - \rho)}$	$3.7\sqrt{d(\rho_s - \rho)}$
(5) μ'_e (Bingham model)	$\dfrac{\tau_s}{\tau_s - \tau_0}\mu_p$	$\dfrac{\tau_s}{\tau_s - \tau_0}\mu_p$	$\dfrac{\tau_s}{\tau_s - \tau_0}\mu_p$

Table XXIV (end)

LIFT OF CUTTINGS IN THE ANNULUS

Parameters to be calculated	SI units	American units	Other units
(6) μ'_e (Power-law model)	$\dfrac{\tau_s}{\left(\dfrac{\tau_s}{K}\right)^{1/n}}$	$479\,\dfrac{\tau_s}{\left(\dfrac{\tau_s}{K}\right)^{1/n}}$	$479\,\dfrac{\tau_s}{\left(\dfrac{\tau_s}{K}\right)^{1/n}}$
(7) Re_s	$\dfrac{V_s \varphi_p \rho}{\mu}$	$\dfrac{15.46 V_s \varphi_p \rho}{\mu}$	$\dfrac{16.67 V_s \varphi_p \rho}{\mu}$
(8) F_L	$1 + \dfrac{\varphi_p}{D_o - D_i}$	$1 + \dfrac{\varphi_p}{D_o - D_i}$	$1 + \dfrac{\varphi_p}{25.4(D_o - D_i)}$
(9) F_T	$\dfrac{D_o - D_i - 1.6\varphi_p}{D_o - D_i - \varphi_p}$	$\dfrac{D_o - D_i - 1.6\varphi_p}{D_o - D_i - \varphi_p}$	$\dfrac{2.54(D_o - D_i) - 0.16\varphi_p}{2.54(D_o - D_i) - 0.1\varphi_p}$
(10) V_r	$\dfrac{4Q}{\pi(D_o^2 - D_i^2)} - V_s$	$\dfrac{24.51Q}{D_o^2 - D_i^2} - V_s$	$\dfrac{1.97Q}{D_o^2 - D_i^2} - V_s$

TABLE XXV

SLIP VELOCITY OF CUTTINGS IN THE ANNULUS

PARAMETERS TO BE CALCULATED	SI UNITS	AMERICAN UNITS	OTHER UNITS
(1) V_s General equation.	$44.3 \sqrt{\dfrac{\rho_s - \rho}{\rho} v \dfrac{1}{s} C_r}$	$139 \sqrt{\dfrac{\rho_s - \rho}{\rho} v \dfrac{1}{s} C_r}$	$8.4 \sqrt{\dfrac{\rho_s - \rho}{\rho} v \dfrac{1}{s} C_r}$
(2) V_s Cylinders. Intermediate flow.	$\dfrac{15.23(\rho_s - \rho)^{0.667} \varphi_p}{\rho^{0.333} \mu^{0.333}}$	$\dfrac{174.2(\rho_s - \rho)^{0.667} \varphi_p}{\rho^{0.333} \mu^{0.333}}$	$\dfrac{4.24(\rho_s - \rho)^{0.667} \varphi_p}{\rho^{0.333} \mu^{0.333}}$
(3) V_s Spheres. Turbulent flow.	$53.9 \sqrt{\dfrac{\rho_s - \rho}{\rho} \varphi_p}$	$169 \sqrt{\dfrac{\rho_s - \rho}{\rho} \varphi_p}$	$10.22 \sqrt{\dfrac{\rho_s - \rho}{\rho} \varphi_p}$
(4) V_s Cylinders. Turbulent flow.	$29.45 \sqrt{\dfrac{\rho_s - \rho}{\rho} \varphi_p}$	$92.37 \sqrt{\dfrac{\rho_s - \rho}{\rho} \varphi_p}$	$5.59 \sqrt{\dfrac{\rho_s - \rho}{\rho} \varphi_p}$

TABLE XXVI

HYDRAULIC POWER

PARAMETERS TO BE CALCULATED	SI UNITS	AMERICAN UNITS	OTHER UNITS
(1) $\mathscr{P}_h$	$Q\Delta P$	$\dfrac{Q\Delta P}{1715}$	$\dfrac{Q\Delta P}{442}$
(2) I_h	$\rho Q V = \rho \dfrac{Q^2}{A}$	$\rho \dfrac{Q^2}{6019A}$	$\rho \dfrac{Q^2}{22785A}$
(3) Q_s	$\dfrac{\mathscr{P}_{h_1}}{P_s}$	$\dfrac{1715\mathscr{P}_{h_1}}{P_s}$	$\dfrac{442\mathscr{P}_{h_1}}{P_s}$
(4) ΔP_l	$\dfrac{\mathscr{P}_{h_1}}{Q}$	$\dfrac{1715\mathscr{P}_{h_1}}{Q}$	$\dfrac{442\mathscr{P}_{h_1}}{Q}$
(5) ΔP_c	$\Delta P_i + \Delta P_a + \Delta P_s$	$\Delta P_i + \Delta P_a + \Delta P_s$	$\Delta P_i + \Delta P_a + \Delta P_s$

For ΔP_i, see table XV, Eq. 4; for ΔP_a, see table XVI, Eq. 4; for ΔP_s, see table XVII, equation according to equipment type used)

TABLE XXVI (end)
HYDRAULIC POWER

PARAMETERS TO BE CALCULATED	SI UNITS	AMERICAN UNITS	OTHER UNITS
(6) k_c		$\rho^{0.8}\mu_p^{0.2}\left[\Sigma\left(\dfrac{L}{10141D^{4.8}}\right)+C\right.$ $\left.+\Sigma\left(\dfrac{L}{7952(D_o+D_i)^{1.8}(D_o-D_i)^3}\right)\right]$ $C=2.526\,10^{-4}$ equipment I $C=9.619\,10^{-5}$ equipment II $C=5.335\,10^{-5}$ equipment III $C=4.163\,10^{-5}$ equipment IV	$\rho^{0.8}\mu_p^{0.2}\left[\Sigma\left(\dfrac{L}{90163D^{4.8}}\right)+C\right.$ $\left.+\Sigma\left(\dfrac{L}{7696(D_o+D_i)^{1.8}(D_o-D_i)^3}\right)\right]$ $C=8.79\,10^{-6}$ equipment I $C=3.298\,10^{-6}$ equipment II $C=1.829\,10^{-6}$ equipment III $C=1.427\,10^{-6}$ equipment IV
(7) Q	$\left(\dfrac{JP_s}{k_c}\right)^{1/1.8}$	$\left(\dfrac{JP_s}{k_c}\right)^{1/1.8}$	$\left(\dfrac{JP_s}{k_c}\right)^{1/1.8}$
(8) Q	$\left(\dfrac{0.26\mathscr{P}_{h_1}}{k_c}\right)^{1/2.8}$	$8.834\left(\dfrac{\mathscr{P}_{h_1}}{k_c}\right)^{1/2.8}$	$5.443\left(\dfrac{\mathscr{P}_{h_1}}{k_c}\right)^{1/2.8}$
(9) A (Table XXI)	$\dfrac{Q}{C_e}\sqrt{\dfrac{\rho}{1.975\Delta P_e}}$	$\dfrac{Q}{109C_e}\sqrt{\dfrac{\rho}{\Delta P_e}}$	$\dfrac{Q}{544C_e}\sqrt{\dfrac{\rho}{\Delta P_e}}$

4
Practical Examples

4.1. WORKED EXAMPLES

The examples are given for a hole, a drill string and general equipement, the characteristics of which are specified in paragr. 4.1.1. below.

Using these data, we shall deal with two examples: one for a quasi-Bingham fluid, the other for a quasi-power-law fluid.

All calculations were made with an HP 67 or HP 97 programmable computer.

4.1.1. General data

Motive power available for pumping	$\mathscr{P}_M$	= 1 104 kW	(1 480 HP)
Nominal power at entry to pump	$\mathscr{P}_m$	= 883 kW	(1 184 HP)
Economy factor of pump utilization	F_e	= 0.75	
Surface equipment	Type III		
Service pressure of surface equipment	P_s	= 20 MPa	(200 bar)
Casing 9 5/8 average inner diameter	D_0	= 0.217 m	(8.54 in)
Depth of casing shoe	S	= 2 500 m	
Hole diameter	D_0	= 0.216 m	(8.5 in)
Depth of hole	H	= 3 500 m	
Outer diameter of drill collars	D_i	= 0.159 m	(6.25 in)
Inner diameter of drill collars	D	= 0.073 m	(2.875 in)
Length of drill collars	L	= 200 m	
Outer diameter of drill pipe	D_i	= 0.127 m	(5 in)
Inner diameter of drill pipe	D	= 0.109 m	(4.28 in)
Nozzle coefficient	C_e	= 0.95	
Maximum flow rate limited by Reynolds number	Re	= 1 100	
Minimum velocity in annulus compatible with uplift of cuttings	V_m	= 0.417-0.5 m/s	(25-30 m/min)

4.1.2.   Calculated example for a Bingham fluid

Mud characteristics

Density	ρ	$= 1\,200 \text{ kg/m}^3 \ (1.2 \text{ kg/dm}^3)$
Fann reading at 600 rpm	$\theta_{600} = 60$	
Fann reading at 300 rpm	$\theta_{300} = 38$	
Fann reading at 200 rpm	$\theta_{200} = 30$	
Fann reading at 100 rpm	$\theta_{10} \ = 22$	

4.1.2.1.   Selection of model and determination of rheological parameters

A.   Selection of model.

The calculation is made by the use of equations in Section 2.2.

Deviation from Bingham model   = 0.999733.
Deviation from power-law model = 0.994942.

Therefore we choose the Bingham model.

B.   Determination of rheological parameters

Equations in paragr. 3.2.1.3 are used in the calculations.

Plastic viscosity, $\mu_p = 29.71$ mPa . s (29.71 cP).
Yield-point, $\tau_0 = \ \ 7.08$ Pa   (14.79 lb$_{force}$/100 ft^2).

4.1.2.2.   Determination of maximum flow rate for a Reynolds number of 1 100

Equations in Section 2.3 are used.

The calculation is made for the annulus between drill collars and hole, i.e., 0.159 m and 0.216 m (6.25 in an 8.5 in).
We find $Q_{max} = 0.01795$ m^3/s (1 077 l/min).

4.1.2.3.   Hydraulic program

The calculation is made for a depth of 3 500 m (bottom of hole).

NOTE.   Pressure losses are calculated for turbulent flow.

A.   Maximum power at bit

The calculation is made according to the flowchart No 4, paragr. 3.7.6.

$$Q = Q_l = 0.01795 \text{ m}^3/\text{s} \ (1\,077 \text{ l/min})$$

The calculated diameter of three equal nozzles is 0.007 m (8.84/32 in); in practice, three 9/32 in nozzles may be used.

B.   Maximum hydraulic impact

The calculation is made according to the flowchart No 5, paragr. 3.7.6.
We find, as before, 0.01795 m^3/s (1 077 l/min). The nozzle pattern is also identical.
In this particular case, irrespective of the criterion adopted in the calculations, optimization of the distribution of hydraulic power between the circuit outside the bit nozzles and the bit is not possible because the flow rate can only be Q_l. The pressure loss ΔP_c follows from this flow rate, and ΔP_e is given by $\Delta P_e = \Delta P_l - \Delta P_c$.
Q_l is smaller than Q_s (Q_s = 0.02387 m^3/s (1 432 l/min)); the overall permissible pressure loss ΔP_l is equal to the service pressure of the installation.

4.1.2.4.   Pressure losses

The calculation is made in accordance with the flowchart in Section 3.3, viz.:

(a) For a flow rate Q = 0.0179 m^3/s (1 075 l/min).
(b) For three 9/32 in nozzles.

We find

Pressure losses at the bit ΔP_e = 14.95 MPa (149.47 bar)
Overall pressure loss ΔP = 19.46 MPa (194.6 bar)

4.1.2.5.   Equivalent circulating density

Equations in Table XXIII are used in the calculation. We obtained:

A.   Without allowance for cuttings carried in the annulus

Pressure loss in the annulus ΔP_a = 1.544 MPa (15.44 bar)
Equivalent circulating density ECD = 1 240 kg/m^3 (1.24 kg/dm^3)

B.   With allowance for cuttings carried in the annulus

For a drilling rate of Av = 10 m/h:

Pressure loss in the annulus ΔP_a = 1.544 MPa (15.44 bar)
Equivalent circulating density ECD = 1 250 kg/m^3 (1.25 kg/dm^3)

4.1.2.6.   Velocity and time of uplift of cuttings

The calculation is made according to the flowchart No 2 in Section 3.6.

We shall assume that the diameter of the cuttings is φ_p = 5 mm and that their density is 2 500 kg/m^3 (2.5 kg/dm^3).

Uplift velocity in annulus drill collars/hole V_r = 0.967 m/s (58.04 m/min)
Uplift velocity in annulus drill pipe/hole V_r = 0.664 m/s (39.84 m/min)
Uplift velocity in annulus drill pipe/casing V_r = 0.654 m/s (39.23 m/min)

Time taken by cuttings to be lifted from a depth of
3 500 m to the surface T_r = 87 min

Time taken by mud to be lifted from a depth of
3 500 m to the surface T_{rb} = 77 min

4.1.2.7. Drill pipe running or pulling speed for equivalent pipe moving density (EPMD) equal to equivalent circulating density (ECD)

The calculation is made by use of the flowchart No 1 of Section 3.4, assuming a closed drill-string.

For an equivalent circulating density, ECD = 1 250 kg/m^3, the running time for a length of 27 m is T_g = 33.54 s.

4.1.2.8. Operational and inertial surge and swab pressures

The acceleration or deceleration in m/s^2 will be taken as equal to the pipe running or pulling speed in m/s (T_g = 34 s).
We obtain:

(a) Surge or swab pressure with the bit at the bottom of the hole:
. Running or pulling (flowchart No 1)

$$P_r = \pm 1.916 \text{ MPa} (\pm 19.16 \text{ bar})$$

. Inertial (Table XXII, Eq. No 2)

$$\Delta P_f = \pm 1.87 \text{ MPa} (\pm 18.7 \text{ bar})$$

(b) Equivalent pipe moving density   EPMD $(+)$ = 1 255 kg/m^3 (1.255 kg/dm^3)
                                       EPMD $(-)$ = 1 145 kg/m^3 (1.145 kg/dm^3)

(c) Equivalent inertial density   EIPMD $(+)$ = 1 253 kg/m^3 (1.253 kg/dm^3)
                                   EIPMD $(-)$ = 1 149 kg/m^3 (1.147 kg/dm^3)

4.1.2.9. Gel breakup surge pressure on restarting circulation

Equation 1 of Table XXII is used in the calculation.
For an aged gel: g_t = 47.9 Pa (100 lb$_{force}$/100 ft^2), the surge pressure is P_r = 7.72 MPa (77.2 bar).

4.1.3.   Calculated example for a power-law fluid

Characteristics of mud

Density	ρ = 1 200 kg/m^3 (1.2 kg/dm^3)
Fann reading at 600 rpm	θ_{600} = 60
Fann reading at 300 rpm	θ_{300} = 38
Fann reading at 200 rpm	θ_{200} = 28
Fann reading at 100 rpm	θ_{100} = 18

4.1.3.1.   Selection of model and determination of rheological parameters

A.   *Selection of model*

The equation in Section 2.2 is used in the calculation.

> Deviation from Bingham model   = 0.996452
> Deviation from power-law model = 0.999630

Accordingly we choose the power-law model.

B.   *Determination of rheological parameters*

Equations in paragr. 3.2.1.3 are used.

> Index of rheological behavior $n = 0.68$
> Consistency index $K = 0.27$ Pa . $s^{0.68}$
> $(0.56\,lb_{force} . s^{0.68}/100\,ft^2)$

4.1.3.2.   Determination of maximum flow rate for a Reynolds number of 1 100

Equations in Section 2.3 are used.

The calculation is made for the annulus between drill collars and hole, i.e., 0.159 m and 0.216 m (6.25 in-8.5 in).

The calculated maximum flow rate $Q_{max} = 0.01707$ m^3/s (1 204 l/min).

4.1.3.3.   Hydraulic program

The calculation is made for a depth of 3 500 m (bottom of hole). Pressure losses are computed for a **Bingham** fluid in turbulent flow.

A.   *Maximum power at bit*

The calculation is made according to the flowchart No 4, paragr. 3.7.6.

We obtain $Q = Q_l = 0.01707$ m^3/s (1 024 l/min.).

The calculated diameter of three equal nozzles is 0.0068 m (8.58/32 in); in practice, two 9/32 in nozzles and one 8/32 in nozzle can be used.

B.   *Maximum hydraulic impact*

The calculation is made according to the flowchart No 5, paragr. 3.7.6.

We find, as before, $Q = Q_l = 0.01707$ m^3/s (1 024 l/min). The nozzle pattern is also the same.

As in the example given for the Bingham fluid (paragr, 4.1.2.3), optimization of the distribution of the hydraulic power between the circuit outside the bit and the bit is not possible, because the flow rate can only be Q_l.

Since Q_l is less than Q_s ($Q_s = 0.02387$ m^3/s (1 432 l/min), ΔP_l will always equal P_s.

4.1.3.4. Pressure losses

The flowchart in Section 3.3 is used in the calculation. The calculation is made for:

(a) flow rate $Q = 0.017$ m^3/s (1 020 l/min),
(b) two 9/32 in nozzles and one 8/32 in nozzle.

We find
>Pressure loss at the bit $\Delta P_e = 15.56$ MPa (155.57 bar).
>Total pressure loss $\Delta P\ = 19.15$ MPa (191.55 bar)

4.1.3.5. Equivalent circulating density

Equations of Table XXIII are used in the calculations. We obtain:

A. *Without allowance for cuttings carried in the annulus*

>Pressure loss in the annulus $\Delta P_a\ = 1.087$ MPa (10.87 bar)
>Equivalent circulating density $ECD = 1\ 230$ kg/m^3 (1.23 kg/dm^3)

B. *With allowance for cuttings carried in the annulus*

At a drilling rate $Av = 10$ m/h:

>Pressure loss in annulus $\Delta P_a\ = 1.087$ MPa (10.87 bar)
>Equivalent circulating density $ECD = 1\ 240$ kg/m^3 (1.24 kg/dm^3)

4.1.3.6. Velocity and time of uplift of cuttings

The flowchart No 3 in Section 3.6 is used in the calculations, which are made for cuttings having an equivalent diameter of $\varphi_p = 5$ mm, and a density of 2 500 kg/m^3 (2.5 kg/dm^3).

Uplift velocity in drill collars/hole annulus	$V_r = 0.906$ m/s (54.38 m/min)
Uplift velocity in drill pipe/hole annulus	$V_r = 0{,}591$ m/s (35.46 m/min)
Uplift velocity in drill pipe/casing annulus	$V_r = 0.587$ m/s (35.21 m/min)
Time taken by cuttings to be lifted from a depth of 3 500 m to the surface	$T_r = 97$ min
Time taken by mud to be lifted from a depth of 3 500 m to the surface	$T_{rb} = 82$ min

4.1.3.7. Drill pipe running or pulling speed for equivalent pipe moving density (EPMD) equal to equivalent circulating density (ECD)

The flowchart No 1 in Section 3.4 is used in the calculation, assuming a closed drill string.

For $ECD = 1\ 240$ kg/m^3 the running time for a 27 m length is $T_g = 39.39$ s.

4.1.3.8. Operational and inertial surge and swab pressures

We shall take an acceleration or a deceleration (in m/s^2) equal to the operating speed ($T_g = 40$ s), and shall assume a closed drill string. We obtain:

(a) Surge or swab pressure with the bit at the bottom of the hole:
. Running or pulling P_r = $\pm$ 1.385 MPa (13.85 bar)
    (calculated by flowchart No 1 paragr. 3.4).

   . Inertial ΔP_f= $\pm$ 1.59 MPa ($\pm$ 15.90 bar).
      (calculated by Table XXII, Eq. 2).

(b) Equivalent pipe moving density EPMD $(+)$ = 1 240 kg/m^3 (1.24 kg/dm^3)
                                    EPMD $(-)$ = 1 160 kg/m^3 (1.16 kg/dm^3)

(c) Equivalent inertial density   EIPMD $(+)$ = 1 245 kg/m^3 (1.245 kg/dm^3)
                                  EIPMD $(-)$ = 1 155 kg/m^3 (1.155 kg/dm^3)

4.1.3.9. Gel breakup surge pressure on restarting circulation

Equation 1 of Table XXII is used.
The result is the same as for a Bingham fluid: P_r = 7.72 MPa (77.2 bar) for g_t = 47.9 Pa (100 lb$_{force}$/100 ft^2).

4.2 NOMOGRAMS

Nomograms can be used in calculations. As examples, some of them are illustrated for the determination of the following parameters:

• Equivalent viscosity of a Bingham fluid (nomogram 1).
• Equivalent viscosity of a power-law fluid (nomogram 2).
• Gel breakup surge pressure (nomogram 3).
• Equivalent mud velocity during drill pipe movement with closed drill string (nomogram 4).
• Surge pressure due to inertial effects (nomogram 5).
• Slip velocity of a cylindrical cutting with an equivalent diameter of 5 mm (nomogram 6).

NOMOGRAM 1. — DETERMINATION OF EQUIVALENT VISCOSITY OF A BINGHAM FLUID

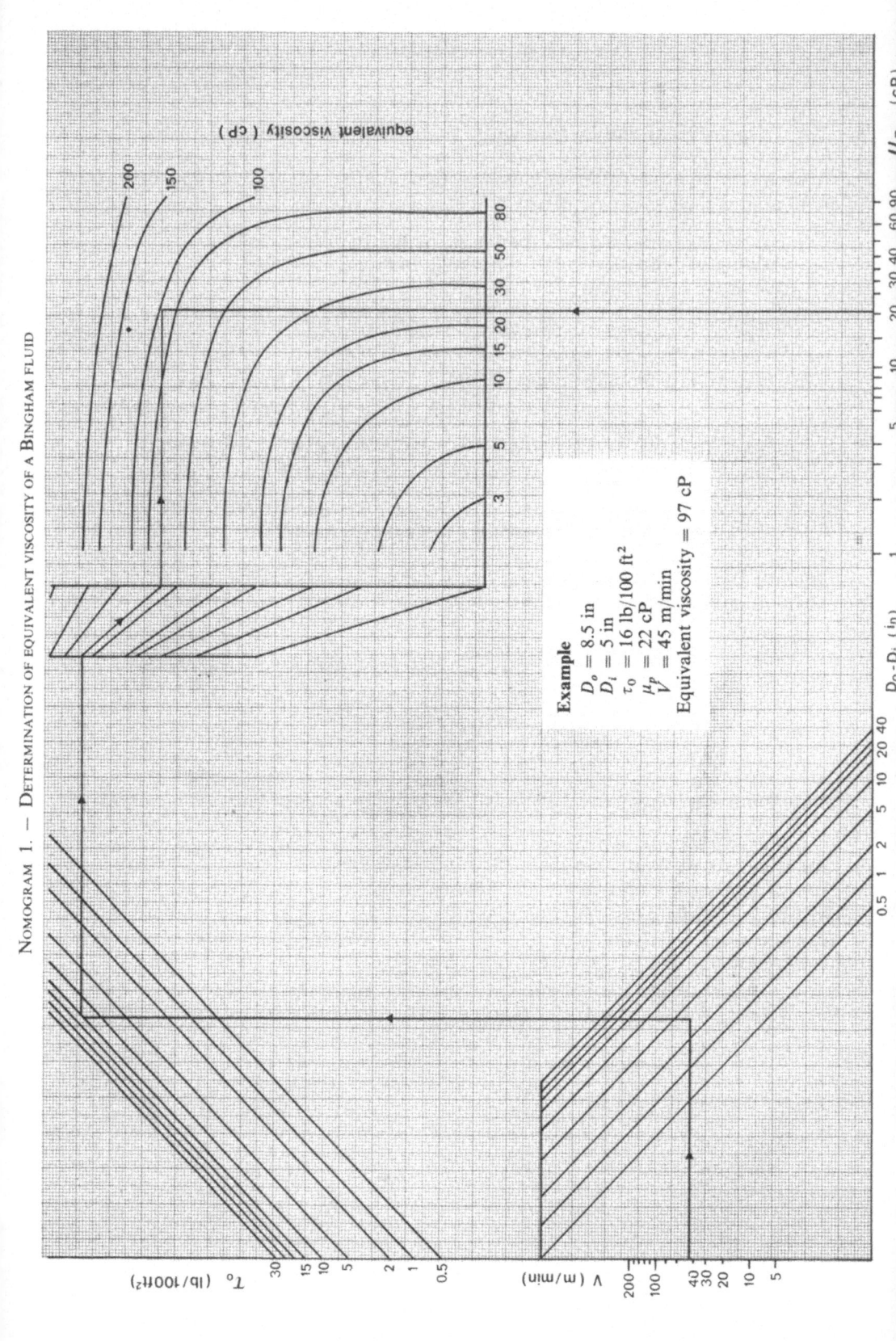

Example

$D_o = 8.5$ in
$D_i = 5$ in
$\tau_o = 16$ lb/100 ft^2
$\mu_p = 22$ cP
$V = 45$ m/min

Equivalent viscosity = 97 cP

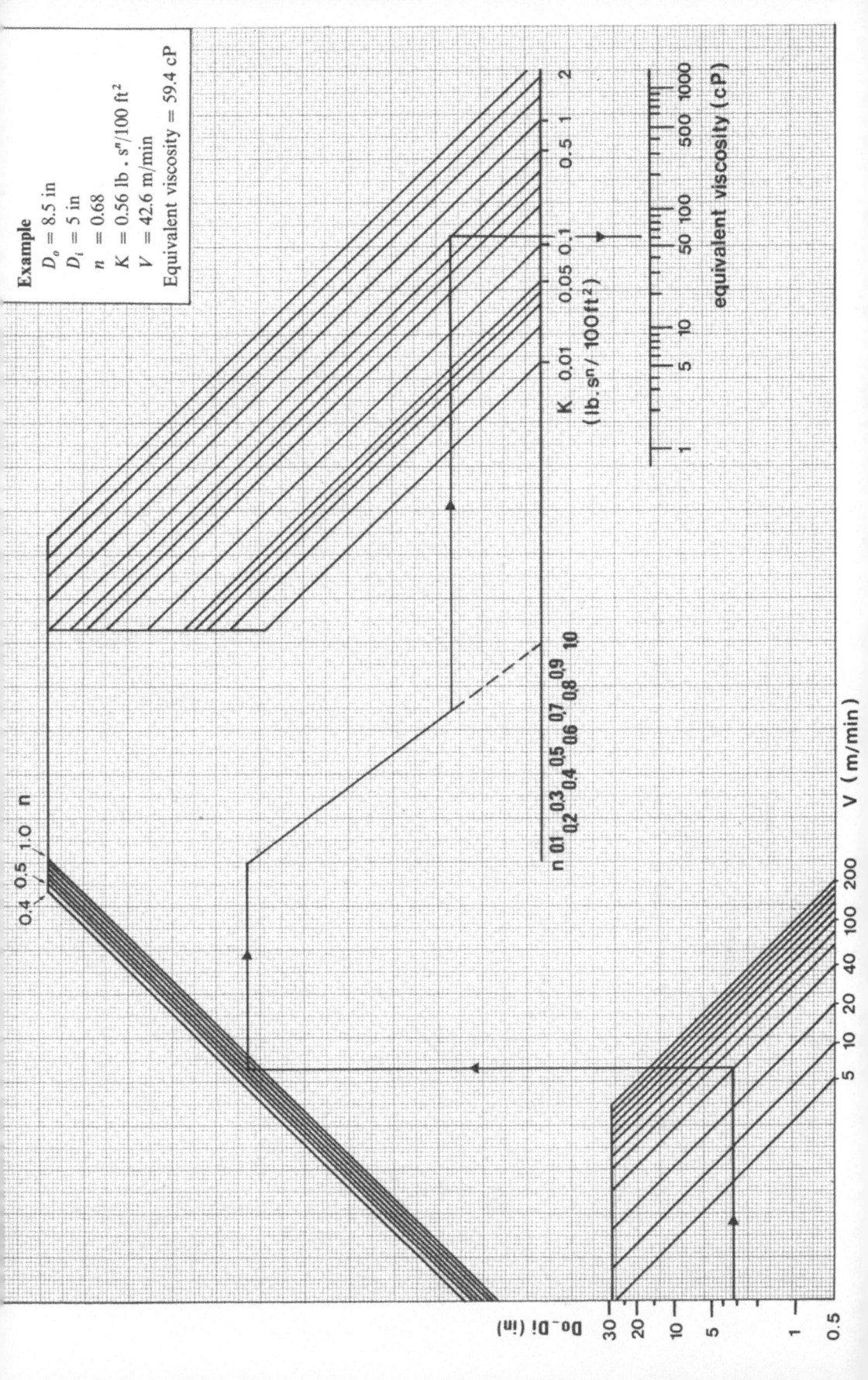

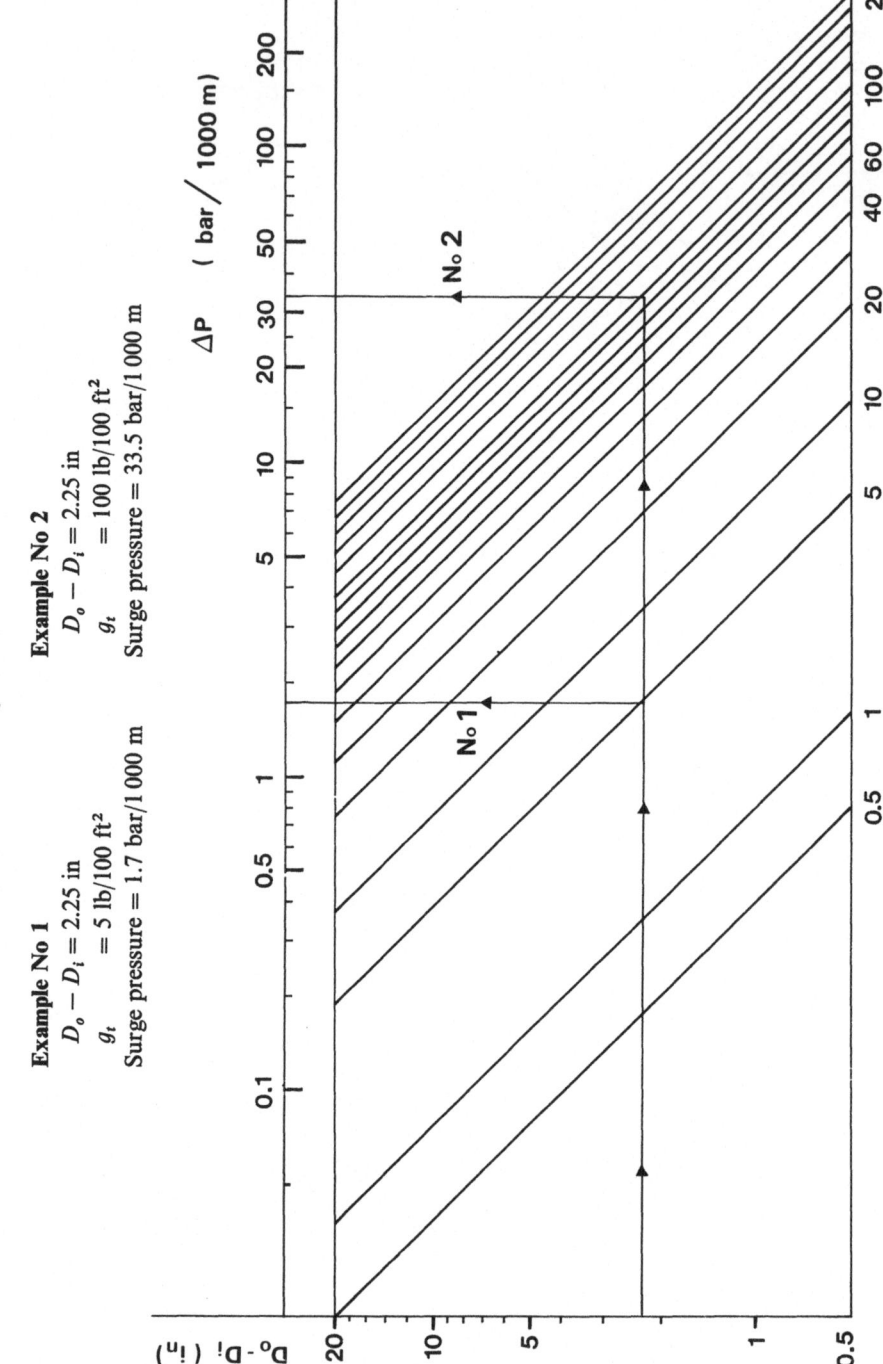

NOMOGRAM 3
SURGE PRESSURE REQUIRED TO BREAK UP THE GEL

Example No 1
$D_o - D_i = 2.25$ in
$g_t \quad = 5$ lb/100 ft²
Surge pressure = 1.7 bar/1 000 m

Example No 2
$D_o - D_i = 2.25$ in
$g_t \quad = 100$ lb/100 ft²
Surge pressure = 33.5 bar/1 000 m

$\Delta P \quad ($ bar $/ 1000$ m $)$

Gel strength   (lb/100ft²)

$D_o - D_i$ (in)

EQUIVALENT VELOCITY OF THE MUD DURING DRILL PIPE MOVEMENT (CLOSED DRILL STRING)

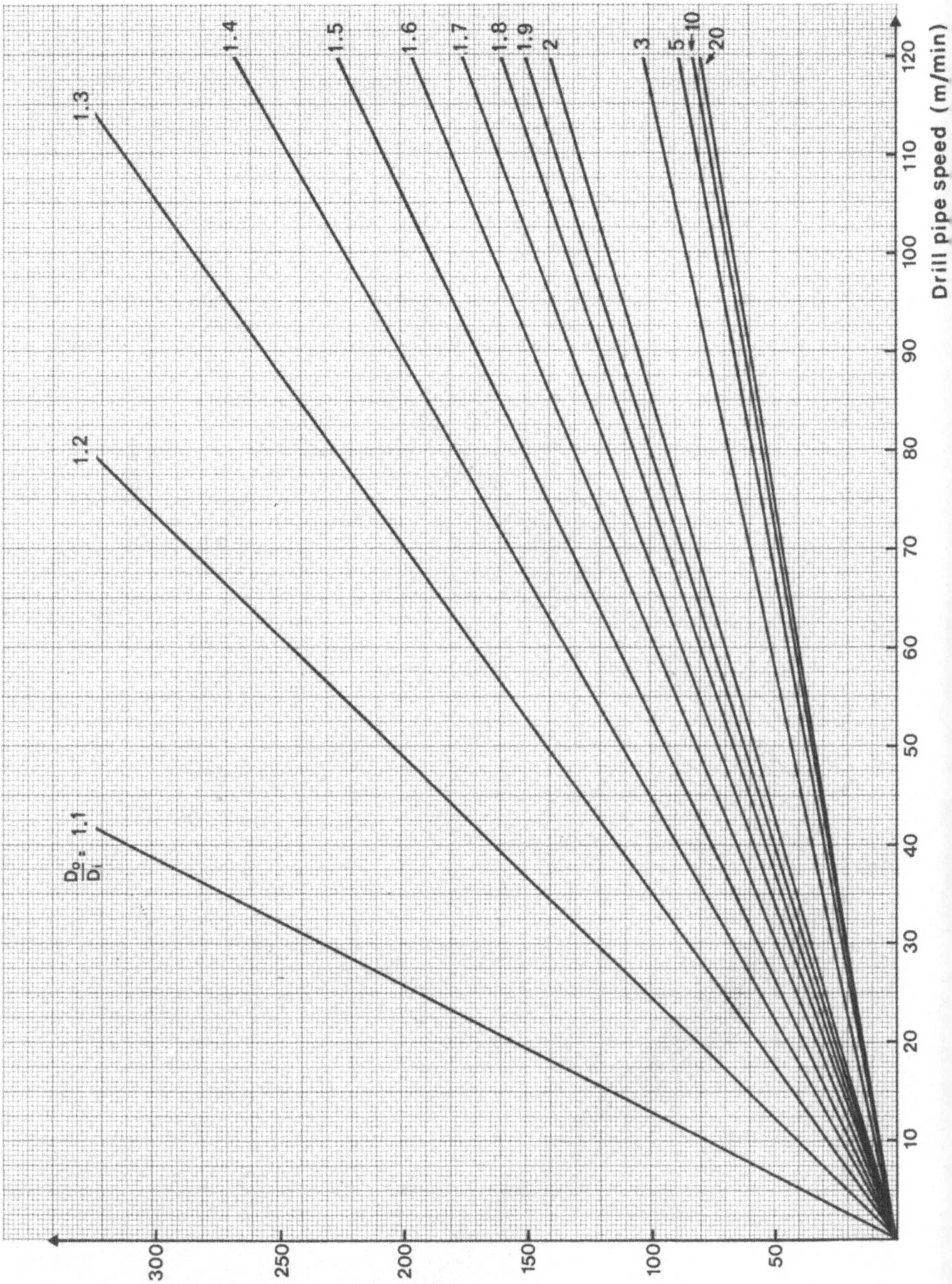

Drill pipe speed (m/min)

Equivalent velocity (m/min)

300

250

200

150

100

50

10   20   30   40   50   60   70   80   90   100   110   120

$\frac{D_o}{D_i} = 1.1$

1.2

1.3

1.4

1.5

1.6

1.7

1.8

1.9

2

3

5

10

20

Nomogram 5
Surge pressure due to inertial effects (closed drill string)

Example No 1
$D_o = 8.5$ in
$D_i = 5$ in
$\rho = 1.2$
$a_p = 0.79$ m/s^2
Surge pressure = 0.5 bar/100 m

Example No 2
$D_o = 10\ 5/8$ in
$D_i = 5$ in
$\rho = 1.6$
$a_p = 5$ m/s^2
Surge pressure = 2.3 bar/100 m

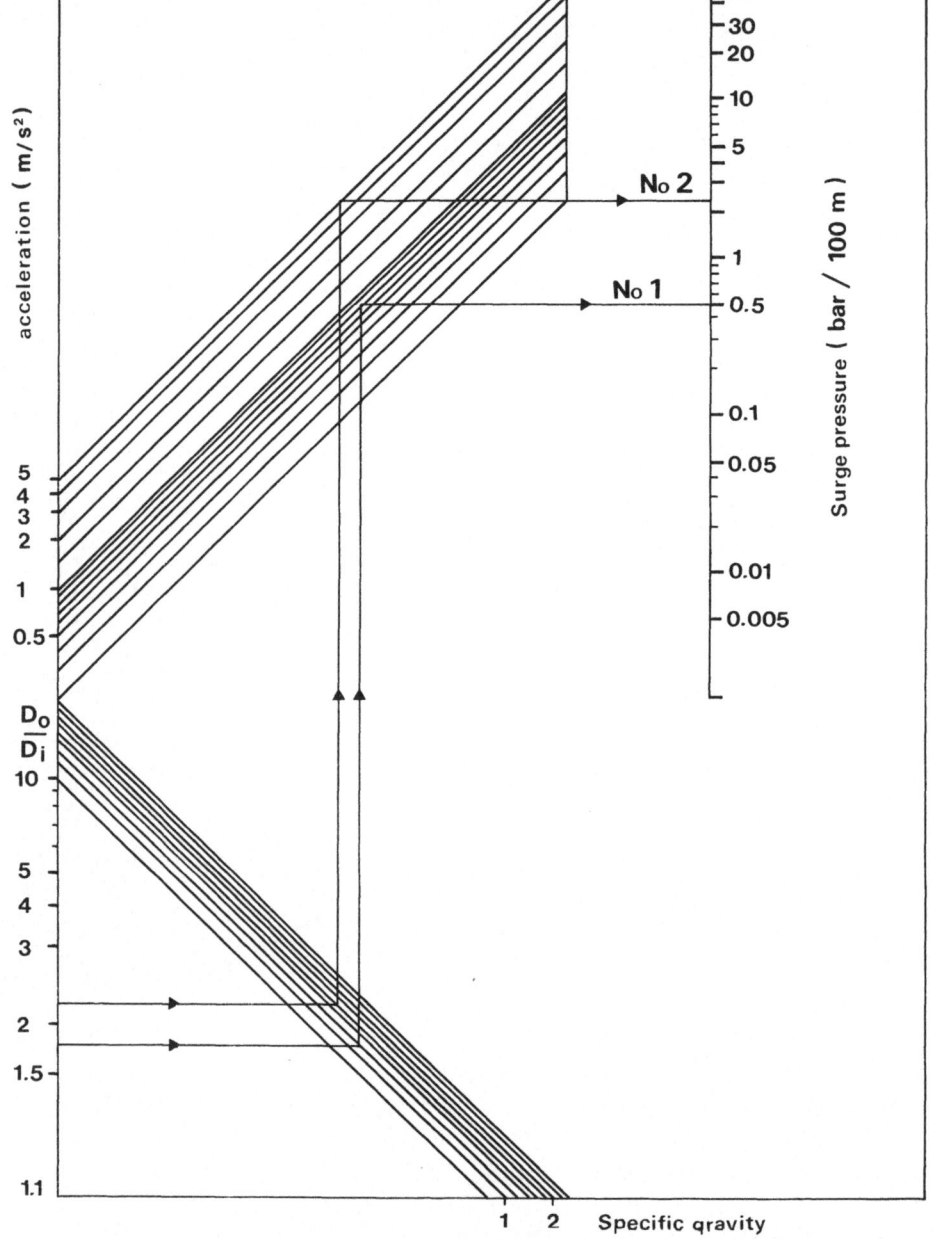

NOMOGRAM 6
SLIP VELOCITY OF A CYLINDRICAL CUTTING
WITH EQUIVALENT DIAMETER OF 5 mm AND A DENSITY OF 2 500 kg/m³

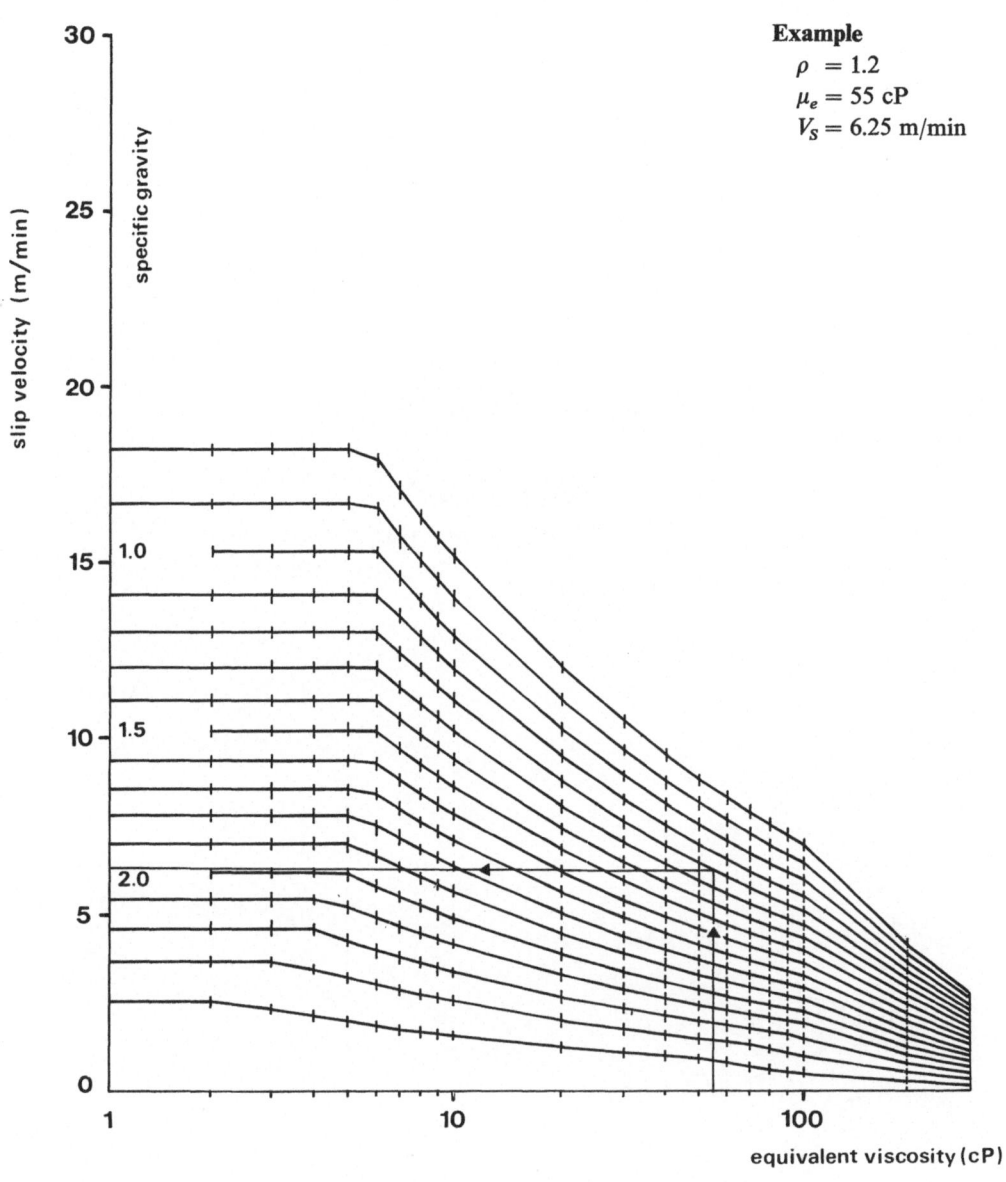

Example
$\rho = 1.2$
$\mu_e = 55$ cP
$V_S = 6.25$ m/min

ACHEVÉ D'IMPRIMER
SUR LES PRESSES
DE L'IMPRIMERIE NOUVELLE
45000 ORLÉANS
EN MAI 1982
N° d'impression: 8489
N° d'Éditeur: 597

Dépôt légal: 3e trimestre 1982.